Adaptive Detection for Multichannel Signals in Non-Ideal Environments

This book systematically presents adaptive multichannel signal detection in three types of non-ideal environments, including sample-starved scenarios, signal mismatch scenarios, and noise plus subspace interference environments.

The authors provide definitions of key concepts, detailed derivations of adaptive multichannel signal detectors, and specific examples for each non-ideal environment. In addition, the possible future trend of adaptive detection methods is discussed, as well as two further research points – namely, the adaptive detection algorithms based on information geometry, and the hybrid approaches that combine adaptive detection algorithms with machine learning algorithms.

The book will be of interest to researchers, advanced undergraduates, and graduate students in sonar, radar signal processing, and communications engineering.

Zeyu Wang is an associate research fellow at the School of Information and Communication Engineering, Beijing University of Posts and Telecommunications. Her research interests include statistical signal processing and radar target detection. She has authored and co-authored more than 20 scientific publications in international journals and conferences.

Weijian Liu is an associate professor at Wuhan Electronic Information Institute, Wuhan, China. His research interests include multichannel signal detection and statistical and array signal processing. He is an editorial board member for several international journals, including *IEEE Transactions on Aerospace and Electronic Systems*, *Signal Processing*, and *Digital Signal Processing*.

Hongmeng Chen (Senior Member, IEEE) is a senior engineer at the Beijing Institute of Radio Measurement, Beijing. His research interests include synthetic aperture radar (SAR)/inverse SAR (ISAR) imaging, forward-looking radar (FLR) imaging, ground moving target indication (GMTI), and motion parameter estimation. He has authored and co-authored more than 50 scientific publications in international journals and conferences. He is an editorial board member for several international journals, including *Journal of Radars* and *Aero Weaponry*. He is also an excellent reviewer of the journal *Digital Signal Processing*.

Adaptive Detection for Multichannel Signals in Non-Ideal Environments

Zeyu Wang, Weijian Liu, and
Hongmeng Chen

CRC Press
Taylor & Francis Group
Boca Raton London New York

CRC Press is an imprint of the
Taylor & Francis Group, an **informa** business

Designed cover image: © mr. teerapon tiuekhomm

This book is published with financial support from the National Natural Science Foundation of China under Grants 62301073 and 62071482, and the Foundation of National Key Laboratory of Radar Signal Processing under Grant JKW202301.

First edition published 2024
by CRC Press
2385 NW Executive Center Drive, Suite 320, Boca Raton FL 33431

and by CRC Press
4 Park Square, Milton Park, Abingdon, Oxon, OX14 4RN
CRC Press is an imprint of Taylor & Francis Group, LLC

ISBN: 978-1-032-76292-0 (hbk)
ISBN: 978-1-032-76961-5 (pbk)
ISBN: 978-1-003-47790-7 (ebk)

DOI: 10.1201/9781003477907

Typeset in Minion
by Deanta Global Publishing Services, Chennai, India

Contents

Acronyms

ABORT	Adaptive Beamformer Orthogonal Rejection Test
ACE	Adaptive Coherence Estimator
AMF	Adaptive Matched Filter
AR	Autoregressive
ASD	Adaptive Subspace Detector
CUT	Cell Under Test
CFAR	Constant False Alarm Rate
CG	Compound Gaussian
CNN	Convolutional Neural Network
CNR	Clutter-to-Noise Ratio
DOF	Degree of Freedom
FIM	Fisher Information Matrix
GLRT	Generalized Likelihood Ratio Test
IID	Independent and Identically Distributed
INR	Interference-to-Noise Ratio
IPIX	Intelligent Pixel Processing X-band
KNN	K-Nearest Neighbor
MAP	Maximum A Posteriori
MF	Matched Filter
MIMO	Multiple-Input Multiple-Output
MLE	Maximum Likelihood Estimate
NMF	Normalized Matched Filter
PCA	Principal Component Analysis
PD	Probability of Detection
PDF	Probability Density Function
PFA	Probability of False Alarm
PHE	Partially Homogeneous Environment
RMB	Reed-Mallet-Brennan
SAR	Synthetic Aperture Radar

SCNR	Signal-to-Clutter-plus-Noise Ratio
SINR	Signal to Interference and Noise Ratio
SIRV	Spherically Invariant Random Vector
SNR	Signal-to-Noise Ratio
UMP	Uniformly Most Powerful
VSL	Value of Significance Level

Symbols

$(\bullet)^*$	The complex conjugate operator
$(\bullet)^T$	The transpose operator
$(\bullet)^H$	The complex conjugate transpose operator
$E(\bullet)$	The statistical expectation
$\mathbf{det}(\bullet)$	The determinant
$\mathbf{tr}(\bullet)$	The trace
In(•)	The logarithm
I_m	The m-dimensional identity matrix
$\mathbf{Re}(\bullet)$	The real part of a complex number
$\mathbf{Im}(\bullet)$	The imaginary part of a complex number
$\mathbf{exp}(\bullet)$	The exponential function
$\mathbb{R}^{N\times L}$	$N\times L$-Dimensional real matrix
$\mathbb{C}^{N\times L}$	$N\times L$-Dimensional complex matrix
$\otimes$	The Kronecker product
$\delta(\bullet)$	The Kronecker delta function
$\mathbf{vec}(\bullet)$	The vectorization operator
$\|\bullet\|$	The Frobenius norm
$\lvert\cdot\rvert$	The modulus of a complex number
$\boldsymbol{y} \sim \quad {}_{N}(\boldsymbol{\mu}, \boldsymbol{R})$	The random vector $\boldsymbol{y}$ follows an $N\times L$-dimensional complex Gaussian distribution with a mean μ and a covariance matrix $\boldsymbol{R}$
$z \sim \chi_N^2(\rho)$	The random variable z follows a noncentral complex Chi-square distribution with N DOFs and a noncentrality parameter ρ
$t \sim \quad {}_{K,M}(\rho)$	The random variable t follows a noncentral complex Beta distribution with K and M DOFs, and a noncentrality parameter ρ

$f \sim \quad {}_{K,M}(\rho)$	The random variable f follows a noncentral complex F distribution with K and M DOFs, and a noncentrality parameter ρ

CHAPTER 1

Overview of Adaptive Detection

1.1 INTRODUCTION

With the increase in computation power and advances in hardware design, the received signals of the sensors are usually multichannel data. Taking radar as an example, with the application of pulse-Doppler techniques and transmit/receive module, radar echoes usually include multi-pulse data in the time domain and multi-antenna data in the space domain. Moreover, the received signals will also be multichannel when the polarization diversity, frequency diversity, and waveform diversity are applied. In other fields like image processing, the hyperspectral image contains dozens to hundreds of spectral channels and each pixel of the hyperspectral image corresponds to a spectral curve, which can describe fine spectral characteristics of land covers. Compared with the single-channel signals, multichannel signals provide more comprehensive information.

Adaptive detection for multichannel signals has become a hot research topic in recent decades in the fields of radar, sonar, communication, speech, image, and so on. It is generally believed that the pioneering work about the adaptive detection for multichannel signals is Kelly's generalized likelihood ratio test (GLRT) [1]. Kelly's GLRT focuses on multichannel radar signal detection in Gaussian noise. In particular, the problem of determining whether the received echoes contain the signal of interest is expressed as binary hypothesis testing. The null hypothesis indicates

DOI: 10.1201/9781003477907-1

that the received echoes include only noise. The alternative hypothesis indicates that the signal of interest is also included in the received echoes besides the noise. The training data, which are independent and identically distributed (IID) with the data under test (also called the primary data) and contain no target signal, are also assumed to be available. The joint probabilities density functions (PDFs) of the training data and data under test are maximized under both hypotheses with respect to the unknown parameters and the GLRT detection statistic, in the form of the ratio of the maxima, is compared with the detection threshold to make a decision.

Adaptive detection for multichannel signals can be classified as adaptive point-like target detection and adaptive distributed target detection according to the extension of a target [1–32]. The range cell the target occupied is one for the point-like target and more than one for the distributed target. The point-like target model is considered in Kelly's GLRT. In Kelly's GLRT, the target signal is rank-one, namely, the target signal is the product of a known vector and an unknown scaling factor. Under such an assumption, the GLRT is not optimal and a uniformly most powerful (UMP) test does not exist for the problem. To investigate detection strategies that may have better detection performance or less computational complexity, the adaptive matched filter (AMF) is derived in a two-step way [2]. In the first step, the GLRT test statistic is derived under the condition that the noise covariance matrix is known. In the second step, the estimated covariance matrix based on the training data is inserted into the test statistic in place of the true one. The AMF achieves a higher probability of detection (PD) than the GLRT for a high signal-to-noise ratio (SNR). Other detectors like the Rao test and Wald test [3, 4] are also derived for adaptive detection of rank-one signals in the Gaussian noise. The Wald test is shown to be the AMF although the derivation is different.

The signal of interest is multi-rank in applications such as polarimetric target detection [5–7], multiuser detection [8, 9], and signal estimation in multipath environments [10]. For polarimetric target detection [5], the target signal is the product of a known multi-rank matrix and an unknown vector. The known multi-rank matrix is constituted by the vectors for different polarization channels, each of which is related to the target signal in the space-time domain. The unknown vector is constituted by the unknown phase, polarization, and amplitude parameters of the target signal for different polarization channels. For multiuser detection [8], the received signal filtered by a chip-matched filter and sampled at the chip rate is the product of a full column rank matrix related to the normalized signalling waveform of users and a vector related to the unknown amplitude and symbol stream. For signal estimation in multipath environments [10], the observed signal after a matched filter (MF) is described as the product of a matrix and a column vector related to the attenuation.

The resolution of radar is gradually improved with the development of the wideband radar. The point-like target model is no longer valid when detecting targets like large ships. The high-resolution radar resolves the target into scattering centres distributed in different range cells. The range cells the target occupies depend on the target size and radar resolution. When the scattering centres of the target are assumed to share the same steering vector, Conte et al. derived the one-step and two-step GLRT detectors [11]. It is concluded that a significant detection gain can be obtained by increasing the radar resolution capabilities. Shuai et al. derived the Rao and Wald detectors based on the one-step and two-step design strategies to detect the range distributed target in [12]. It is found that the one-step algorithms have the same asymptotic performance as the one-step GLRT, while the two-step algorithms are equivalent to the two-step GLRT. When the signal of interest is only known to belong to a subspace, Besson et al. derived the generalized adaptive direction detector is derived according to the two-step GLRT [13]. When the useful signal is completely unknown, Conte et al. derived detectors based on the one-step GLRT, two-step GLRT, modified two-step GLRT, and the spectral norm [14]. For the same problem, Liu et al. designed the Rao test and Wald test [15]. It is found that the Wald test is identical to the two-step GLRT and the Rao test is equivalent to the modified two-step GLRT.

The above adaptive multichannel signal detectors focus on the scenarios wherein the training data are sufficient enough and there is no signal mismatch. However, the ideal assumption is difficult to satisfy due to the diversified targets and complex environment. In this book, three types of non-ideal environments will be discussed: sample-starved scenarios, signal mismatch scenarios, and adaptive detection in subspace interference plus noise.

1.2 BLOCK DIAGRAM AND SIGNAL MODEL

1.2.1 Block Diagram for Adaptive Detection

The conventional detection algorithm usually includes two procedures: filter and detection. Taking radar as an example, the filter is designed to suppress the noise in the first procedure. In general, the moving target indicator technology is used to design the filter for multichannel data in the time domain, while space-time adaptive processing is used for the multichannel data in the angular and Doppler domains. In the second procedure, the fast Fourier transform algorithm and the cell averaging constant false alarm rate (CFAR) technology are used to detect signals. The aim of the conventional detection algorithm is to maximize the output SNR.

Different from the conventional detection algorithm, the design criterion of adaptive multichannel signals detection is to maximize the PD. The block diagram for adaptive detection is shown in Figure 1.1. Both the primary data and training data are utilized to design adaptive detectors and then the resulting detectors are compared with the detection threshold which is set according to the pre-assigned probability of false alarm (PFA) to make a decision whether the useful signal is present or not.

From Figure 1.1, the filter and target detection are both included in the design of the adaptive detector. In other words, adaptive multichannel signals detection realizes the integration of noise suppression and detection. Compared with the conventional detection algorithm, adaptive multichannel signals detection has the following advantages:

(1) The structure of the adaptive multichannel signals detection algorithms is simpler since the filter and signal detection are both included in the adaptive detector and a separate design of the filter is not required.

(2) The adaptive detectors always ensure the CFAR property. That is to say, the PFA of the adaptive detector is independent of the noise power or the noise covariance matrix. The CFAR property is important to design an effective detector since the PFA may be dramatically raised to an unaffordable value when the noise changes severely and the detector does not have the CFAR property [16].

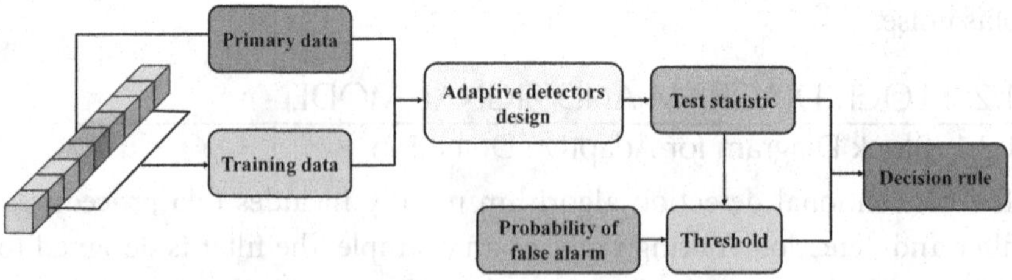

FIGURE 1.1 The block diagram for adaptive detection.

(3) The design criteria used to design adaptive detectors are flexible. Different design criteria such as the GLRT, Rao test, Wald test, and gradient test can be exploited to design adaptive detectors according to the detection problem to be solved.

1.2.2 Multichannel Signal Model

It is supposed that the primary data and the training data are denoted by $\boldsymbol{y}$ and $\boldsymbol{y}_k, k=1,\ldots,K$, where K is the number of training data. The problem of detecting a multichannel signal is formulated as the following binary hypothesis testing:

$$\begin{cases} H_0 : \boldsymbol{y} = \boldsymbol{n}, & \boldsymbol{y}_k = \boldsymbol{n}_k, k = 1,\ldots,K \\ H_1 : \boldsymbol{y} = \boldsymbol{s} + \boldsymbol{n}, & \boldsymbol{y}_k = \boldsymbol{n}_k, k = 1,\ldots,K \end{cases} \tag{1.1}$$

where $\boldsymbol{n}$ and $\boldsymbol{n}_k$ denote the noise including the clutter, thermal noise, and interference, $\boldsymbol{s}$ denotes the useful multichannel signal. $\boldsymbol{s},\boldsymbol{y},\boldsymbol{y}_k,\boldsymbol{n}$ and $\boldsymbol{n}_k$ are all N-dimensional complex vectors. The expressions of the multichannel signal $\boldsymbol{s}$ vary in different applications.

For the pulse-Doppler radar, the received data consists of multiple pulses. In this scenario, the multichannel signal is $\boldsymbol{s} = \alpha \boldsymbol{v}$, where $\boldsymbol{v} = \left[1, e^{j2\pi f_d}, \ldots, e^{j(N-1) f_d}\right]^T$ denotes the temporal steering vector, α denotes the amplitude, N denotes the number of pulses, f_d denotes the normalized Doppler frequency. The temporal multichannel signal data model is shown in Figure 1.2.

For the phased array radar, the received data can be a spatial or spatial-temporal multichannel signal. We take the airborne phased array radar which uses a uniform linear array as an example and set the number of elements of the uniform linear array as N_s. The schematic diagram of the airborne phased array radar system is shown in Figure 1.3. In such a scenario, the multichannel signal is $\boldsymbol{s} = \alpha \boldsymbol{v}$, where $\boldsymbol{v} = \boldsymbol{v}_d \otimes \boldsymbol{v}_s$ denotes the spatial-temporal steering vector, α denotes the amplitude, $\boldsymbol{v}_d = \left[1, e^{j2\pi f_d}, \ldots, e^{j(N_t-1) f_d}\right]^T$, $\boldsymbol{v}_s = \left[1, e^{j2\pi f_s}, \ldots, e^{j(N_s-1) f_s}\right]^T$, f_s denotes the normalized spatial frequency, and N_t denotes the number of pulses. The spatial-temporal multichannel signal model is shown in Figure 1.4.

For the sonar system, a sound wave is used as the transmitted signal. The condition of coherent accumulation is not satisfied for the sonar system due to the long-time delay of the received signal caused by the

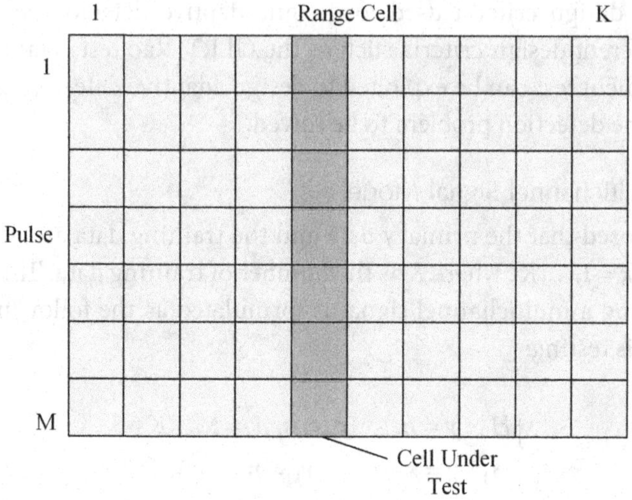

FIGURE 1.2 Temporal multichannel signal model.

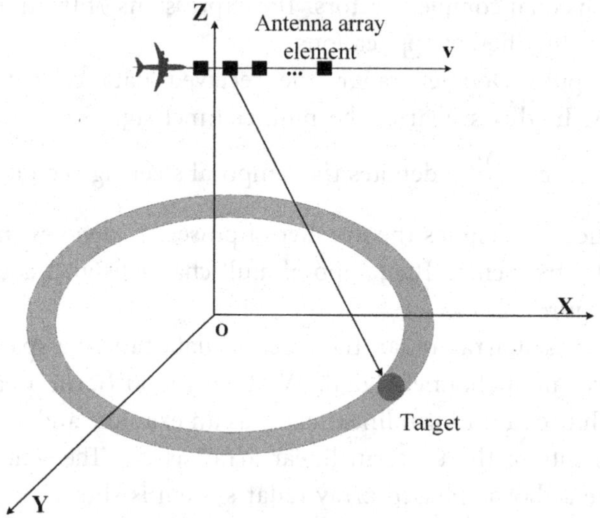

FIGURE 1.3 The airborne phased array radar system.

slow speed of the underwater acoustic signal. Therefore, monopulse mode is usually exploited when the sonar system transmits signals. In this scenario, the multichannel target signal is $\boldsymbol{s} = \alpha \boldsymbol{v}$, where $\boldsymbol{v} = \boldsymbol{v}_t \otimes \boldsymbol{v}_s$ denotes the spatial-temporal steering vector, α denotes the amplitude, $\boldsymbol{v}_d = \left[1, e^{j2\pi f_d}, \ldots, e^{j(N_t-1)f_d}\right]^T$, $\boldsymbol{v}_s = \left[1, e^{j2\pi\kappa f_s}, \ldots, e^{j(N_s-1)\kappa f_s}\right]^T$, κ denotes the correction factor.

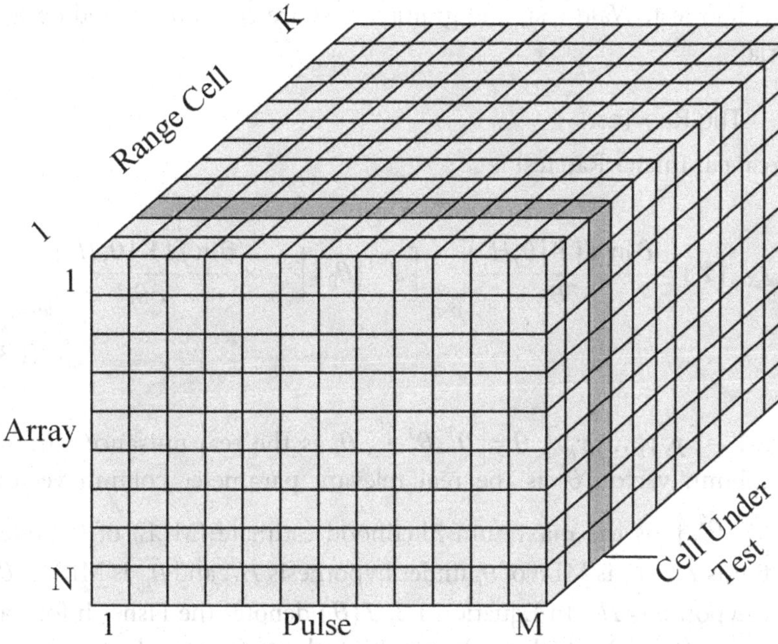

FIGURE 1.4 Spatial-temporal multichannel signal model.

1.3 DESIGN CRITERIA OF ADAPTIVE DETECTORS

When all the parameters related to the target signal and noise are completely known, the likelihood ratio test is a clairvoyant detector. For the detection problem in Equation 1.1, the likelihood ratio test is

$$\Gamma(y, y_1, \ldots y_K) = \frac{f(y, y_1, \ldots, y_K \mid H_1)}{f(y, y_1, \ldots, y_K \mid H_0)} \underset{H_0}{\overset{H_1}{\gtrless}} T_h \tag{1.2}$$

where $f(y, y_1, \ldots y_K \mid H_i)$ denotes the joint PDF of the primary data and training data, the threshold T_h is determined by the equation: $\int_{\{(y, y_1, \ldots, y_K), \Gamma(y, y_1, \ldots y_K) > T_h\}} f(y, y_1, \ldots y_K \mid H_0) = P_{fa}$.

In practical scenarios, the likelihood ratio test is unrealizable since the parameters are usually unknown, such as the amplitude of the signal and the statistical properties of the noise. However, the performance of the likelihood ratio test which assumes perfect knowledge of unknown parameters will be an upper bound. When unknown parameters exist, the

GLRT, Rao test, Wald test, and gradient test are commonly used design criteria.

1.3.1 The Rao Test

The real parameter Rao test is

$$\Gamma_{real\_Rao}(\boldsymbol{Y}) = \left.\frac{\partial \ln f(\boldsymbol{Y}|\boldsymbol{\theta},H_1)}{\partial \boldsymbol{\theta}_r}\right|_{\boldsymbol{\theta}=\hat{\boldsymbol{\theta}}_0}^T \left[\boldsymbol{J}^{-1}(\hat{\boldsymbol{\theta}}_0)\right]_{\boldsymbol{\theta}_r,\boldsymbol{\theta}_r} \times \left.\frac{\partial \ln f(\boldsymbol{Y}|\boldsymbol{\theta},H_1)}{\partial \boldsymbol{\theta}_r}\right|_{\boldsymbol{\theta}=\hat{\boldsymbol{\theta}}_0} \tag{1.3}$$

where $\boldsymbol{Y} = [\boldsymbol{y}, \boldsymbol{y}_1, \ldots, \boldsymbol{y}_K]$, $\boldsymbol{\theta} = [\boldsymbol{\theta}_r^T, \boldsymbol{\theta}_s^T]^T$, $\boldsymbol{\theta}_s$ is the real nuisance parameter column vector, $\boldsymbol{\theta}_r$ is the real relevant parameter column vector, $\hat{\boldsymbol{\theta}}_0 = [\hat{\boldsymbol{\theta}}_{r0}^T, \hat{\boldsymbol{\theta}}_{s0}^T]^T$ is the maximum likelihood estimate (MLE) of $\boldsymbol{\theta}$ under hypothesis H_0, $\hat{\boldsymbol{\theta}}_{r0}$ is MLE of $\boldsymbol{\theta}_r$ under hypothesis H_0, and $\hat{\boldsymbol{\theta}}_{s0}$ is MLE of $\boldsymbol{\theta}_s$ under hypothesis H_0. In Equation 1.3, $\boldsymbol{J}(\boldsymbol{\theta})$ denotes the Fisher information matrix (FIM)), which can be partitioned into four blocks

$$\boldsymbol{J}(\boldsymbol{\theta}) = \begin{bmatrix} \boldsymbol{J}_{\boldsymbol{\theta}_r,\boldsymbol{\theta}_r}(\boldsymbol{\theta}) & \boldsymbol{J}_{\boldsymbol{\theta}_r,\boldsymbol{\theta}_s}(\boldsymbol{\theta}) \\ \boldsymbol{J}_{\boldsymbol{\theta}_s,\boldsymbol{\theta}_r}(\boldsymbol{\theta}) & \boldsymbol{J}_{\boldsymbol{\theta}_s,\boldsymbol{\theta}_s}(\boldsymbol{\theta}) \end{bmatrix} \tag{1.4}$$

where $\boldsymbol{J}_{\boldsymbol{\theta}_r,\boldsymbol{\theta}_r}(\boldsymbol{\theta}) = -E\left[\frac{\partial^2 \ln f(\boldsymbol{Y}|\boldsymbol{\theta})}{\partial \boldsymbol{\theta}_r \partial \boldsymbol{\theta}_r^T}\right]$, $\boldsymbol{J}_{\boldsymbol{\theta}_r,\boldsymbol{\theta}_s}(\boldsymbol{\theta}) = -E\left[\frac{\partial^2 \ln f(\boldsymbol{Y}|\boldsymbol{\theta})}{\partial \boldsymbol{\theta}_r \partial \boldsymbol{\theta}_s^T}\right]$,

$\boldsymbol{J}_{\boldsymbol{\theta}_s,\boldsymbol{\theta}_r}(\boldsymbol{\theta}) = -E\left[\frac{\partial^2 \ln f(\boldsymbol{Y}|\boldsymbol{\theta})}{\partial \boldsymbol{\theta}_s \partial \boldsymbol{\theta}_r^T}\right]$, $\boldsymbol{J}_{\boldsymbol{\theta}_s,\boldsymbol{\theta}_s}(\boldsymbol{\theta}) = -E\left[\frac{\partial^2 \ln f(\boldsymbol{Y}|\boldsymbol{\theta})}{\partial \boldsymbol{\theta}_s \partial \boldsymbol{\theta}_s^T}\right]$,

$\left[\boldsymbol{J}^{-1}(\boldsymbol{\theta})\right]_{\boldsymbol{\theta}_r,\boldsymbol{\theta}_r}$ is the $(\boldsymbol{\theta}_r, \boldsymbol{\theta}_r)$ part of the inverse of $\boldsymbol{J}(\boldsymbol{\theta})$ and can be obtained according to the inverse of the partitioned matrix

$$\left[\boldsymbol{J}^{-1}(\boldsymbol{\theta})\right]_{\boldsymbol{\theta}_r,\boldsymbol{\theta}_r} = \left[\boldsymbol{J}_{\boldsymbol{\theta}_r,\boldsymbol{\theta}_r}(\boldsymbol{\theta}) - \boldsymbol{J}_{\boldsymbol{\theta}_r,\boldsymbol{\theta}_s}(\boldsymbol{\theta}) \boldsymbol{J}_{\boldsymbol{\theta}_s,\boldsymbol{\theta}_s}^{-1}(\boldsymbol{\theta}) \boldsymbol{J}_{\boldsymbol{\theta}_s,\boldsymbol{\theta}_r}(\boldsymbol{\theta})\right]^{-1} \tag{1.5}$$

When the unknown parameters are complex, there are two ways to derive the Rao test. The first way is to divide the complex parameter into real and imaginary parts and then derive the real parameter Rao test. For example, to solve the detection problem with the relevant parameter $\boldsymbol{\theta}_{cr} = \boldsymbol{u}_r + j\boldsymbol{v}_r \in \mathbb{C}^{M\times 1}$ and nuisance parameter $\boldsymbol{\theta}_{cs} = \boldsymbol{u}_s + j\boldsymbol{v}_s \in \mathbb{C}^{N\times 1}$. The real parts and imaginary parts of the complex parameters are used to construct $\boldsymbol{\theta}_s$ and $\boldsymbol{\theta}_r$: $\boldsymbol{\theta}_r = \left[\boldsymbol{u}_r^T, \boldsymbol{v}_r^T\right]^T \in \mathbb{R}^{2M\times 1}$, $\boldsymbol{\theta}_s = \left[\boldsymbol{u}_s^T, \boldsymbol{v}_s^T\right]^T \in \mathbb{R}^{2N\times 1}$, and $\boldsymbol{\theta} = \left[\boldsymbol{\theta}_r^T, \boldsymbol{\theta}_s^T\right]^T \in \mathbb{R}^{2(N+M)\times 1}$. The real parameter Rao test is derived by substituting the above $\boldsymbol{\theta}_s$ and $\boldsymbol{\theta}_r$ into Equation 1.3. The second way is to treat the complex parameter as a whole and derive the complex parameter Rao test. The Rao tests derived in the two ways coincide [33].

The complex parameter Rao test is

$$\Gamma_{complex\_Rao}\left(\boldsymbol{Y}\right) = \left.\frac{\partial \ln f\left(\boldsymbol{Y}|\Theta, H_1\right)}{\partial \Theta_r}\right|_{\Theta=\hat{\Theta}_0}^{T} \left[\boldsymbol{J}^{-1}\left(\hat{\Theta}_0\right)\right]_{\Theta_r\Theta_r} \left.\frac{\partial \ln f\left(\boldsymbol{Y}\mid\Theta, H_1\right)}{\partial \Theta_r^*}\right|_{\Theta=\hat{\Theta}_0} \tag{1.6}$$

Where $\Theta_r = \left[\boldsymbol{\theta}_{cr}^T, \boldsymbol{\theta}_{cr}^H\right]^T$, $\Theta_s = \left[\boldsymbol{\theta}_{cs}^T, \boldsymbol{\theta}_{cs}^H\right]^T$, $\Theta = \left[\Theta_r^T, \Theta_s^T\right]^T$, $\hat{\Theta}_0$ denotes the MLE of Θ under H_0, $(\cdot)^*$ denotes the complex conjugate, $\boldsymbol{J}(\Theta)$ is the FIM for the complex-valued signals:

$$\boldsymbol{J}\left(\Theta\right) = E\left[\frac{\partial \ln f\left(\boldsymbol{Y}|H_1\right)}{\partial \Theta^*}\frac{\partial \ln f\left(\boldsymbol{Y}|H_1\right)}{\partial \Theta^T}\right] = \begin{bmatrix} \boldsymbol{J}_{\Theta_r\Theta_r} & \boldsymbol{J}_{\Theta_r\Theta_s} \\ \boldsymbol{J}_{\Theta_s\Theta_r} & \boldsymbol{J}_{\Theta_s\Theta_s} \end{bmatrix} \tag{1.7}$$

where $\boldsymbol{J}_{\Theta_r\Theta_r} = E\left[\frac{\partial \ln f\left(\boldsymbol{Y}|H_1\right)}{\partial \Theta_r^*}\frac{\partial \ln f\left(\boldsymbol{Y}|H_1\right)}{\partial \Theta_r^T}\right]$,

$\boldsymbol{J}_{\Theta_r\Theta_s} = E\left[\frac{\partial \ln f\left(\boldsymbol{Y}|H_1\right)}{\partial \Theta_r^*}\frac{\partial \ln f\left(\boldsymbol{Y}|H_1\right)}{\partial \Theta_s^T}\right]$,

$\boldsymbol{J}_{\Theta_s\Theta_r} = E\left[\frac{\partial \ln f\left(\boldsymbol{Y}|H_1\right)}{\partial \Theta_s^*}\frac{\partial \ln f\left(\boldsymbol{Y}|H_1\right)}{\partial \Theta_r^T}\right]$, and

$\boldsymbol{J}_{\Theta_s\Theta_s} = E\left[\frac{\partial \ln f\left(\boldsymbol{Y}|H_1\right)}{\partial \Theta_s^*}\frac{\partial \ln f\left(\boldsymbol{Y}|H_1\right)}{\partial \Theta_s^T}\right]$ are four blocks of $\boldsymbol{J}(\Theta)$.

$\left[J^{-1}(\Theta)\right]_{\Theta_r\Theta_r}$ is the (Θ_r,Θ_r) part of the inverse of the matrix $J(\Theta)$:

$$\left[J^{-1}(\Theta)\right]_{\Theta_r\Theta_r}=\left[J_{\Theta_r\Theta_r}(\Theta)-J_{\Theta_r\Theta_s}(\Theta)J^{-1}_{\Theta_s\Theta_s}(\Theta)J_{\Theta_s\Theta_r}(\Theta)\right]^{-1} \quad (1.8)$$

1.3.2 The Wald Test

The real parameter Wald test is

$$\Gamma_{real\_Wald}(Y)=\left(\hat{\theta}_{r1}-\theta_{r0}\right)^T\left(\left[J^{-1}\left(\hat{\theta}_1\right)\right]_{\theta_r,\theta_r}\right)^{-1}\left(\hat{\theta}_{r1}-\theta_{r0}\right) \quad (1.9)$$

where $\hat{\theta}_{r1}$ denotes the MLE of θ_r under hypothesis H_1, θ_{r0} denotes the value of θ_r under hypothesis H_0, $\hat{\theta}_1=\left[\hat{\theta}^T_{r1},\hat{\theta}^T_{s1}\right]^T$ denotes the MLE of θ under hypothesis H_1, and $J(\theta)$ denotes the FIM shown in Equation 1.4.

Similar to the Rao test, the Wald test can be derived by dividing the complex parameters into real and imaginary parts or treating the complex parameters as a single quantity when the unknown parameters are complex.

The complex parameter Wald test is

$$\Gamma_{complex\_Wald}(Y)=\left(\hat{\Theta}_{r1}-\Theta_{r0}\right)^H\left\{\left[J^{-1}\left(\hat{\Theta}_1\right)\right]_{\Theta_r\Theta_r}\right\}^{-1}\left(\hat{\Theta}_{r1}-\Theta_{r0}\right) \quad (1.10)$$

where $\hat{\Theta}_{r1}$ denotes the MLE of Θ_r under H_1, $\hat{\Theta}_1$ denotes the MLE of Θ under H_1, Θ_{r0} denotes the value of Θ_r under H_0, $(\cdot)^H$ denotes the complex conjugate transpose, $J(\Theta)$ denotes the FIM shown in Equation 1.7.

1.3.3 The Gradient Test

The real parameter gradient test is given by

$$\Gamma_{real\_Gradient}(Y)=\left.\frac{\partial \ln f(Y|\theta,H_1)}{\partial\theta_r}\right|^T_{\theta=\hat{\theta}_0}\left(\hat{\theta}_{r1}-\theta_{r0}\right) \quad (1.11)$$

Similar to the Rao and Wald tests, the gradient test can be derived in two ways when the unknown parameters are complex.

The complex parameter gradient test is

$$\Gamma_{complex\_Gradient}(\boldsymbol{Y}) = \left.\frac{\partial \ln f(\boldsymbol{Y} \mid \Theta, H_1)}{\partial \Theta_r^T}\right|_{\Theta=\hat{\Theta}_0} \left(\hat{\Theta}_{r1} - \Theta_{r0}\right) \tag{1.12}$$

1.3.4 The Generalized Likelihood Ratio Test

The GLRT is implemented by replacing the unknown parameters in the joint PDF with their MLEs under both hypotheses, mathematically expressed as

$$\Gamma_{GLRT}(\boldsymbol{Y}) = \frac{f\left(\boldsymbol{Y} \mid \hat{\tilde{\theta}}_1, H_1\right)}{f\left(\boldsymbol{Y} \mid \hat{\tilde{\theta}}_0, H_0\right)} \tag{1.13}$$

where $\tilde{\theta} = \left[\theta_r^T, \theta_s^T\right]^T$ holds when the unknown parameters are real, $\tilde{\theta} = \left[\theta_{cr}^T, \theta_{cs}^T\right]^T$ holds when the unknown parameters are complex, and $\hat{\tilde{\theta}}_1$ and $\hat{\tilde{\theta}}_0$ are MLEs of $\tilde{\theta}$ under hypotheses H_1 and H_0, respectively.

1.4 TYPES OF NON-IDEAL ENVIRONMENTS

In this section, three types of non-ideal environments, sample-starved scenarios, signal mismatch scenarios, and the presence of interference, are discussed and the factors that cause the non-ideal environments are analyzed.

1.4.1 The Sample-Starved Scenarios

According to the well-known Reed-Mallet-Brennan (RMB) rule, the number of training data should be at least twice the system degree of freedom (DOF) to ensure a satisfactory detection performance. The training data are usually collected from the range cells around the data under test and should be IID with the data under test. However, in practical nonhomogeneous noise scenarios or dense target scenarios, this condition is difficult to meet, especially for a radar system with a large number of array elements and/or pulses.

Taking the Mountain Top dataset as an example, to obtain sufficient training data, the number of which is at least twice that of the system DOF, the radar echo data within a range of several hundred kilometres are required to be IID. However, the situation is difficult to satisfy since the practical terrain is complex and variable. When the number of training data is too small, the detection performance of the detector decreases. In many practical applications, there are usually too few training data to form a reliable sample covariance matrix, which is used to estimate the unknown noise covariance matrix. Thus, it is necessary to discuss the adaptive detection problem in sample-starved scenarios.

1.4.2 The Signal Mismatch Scenarios

In the classical adaptive multichannel signal detectors such as Kelly's GLRT and AMF, the steering vector of the target or the target signal signature is usually assumed to be perfectly known. However, non-ideal factors like antenna calibration errors, beam pointing errors, wavefront distortions, and multipath propagation effects are inevitable in a practical environment. When these factors exist, the actual steering vector may deviate from the nominal one. In other words, the signal mismatch occurs. Moreover, the jamming signals received by the radar sidelobe can also cause signal mismatch. The schematic diagram of the signal mismatch scenario is shown in Figure 1.5.

Adaptive detectors can be categorized as mismatch robust detectors, mismatch selective detectors, and tunable adaptive detectors according to the selectivity of the detectors to mismatched signals. When the signal mismatch occurs, the mismatch robust detectors can still achieve high PD, while the PD of mismatch selective detectors decreases rapidly with the increase of the degree of signal mismatch. The tunable detectors can adjust the selectivity to mismatched signals according to different requirements by adjusting the tunable parameters. Adaptive detectors with different sensitivities to mismatched signals are needed for different scenarios. Taking radar as an example, radars in scan mode require mismatch robust detectors, while radars in tracking mode prefer mismatch selective detectors. Therefore, it is necessary to design adaptive mismatch robust detectors, adaptive mismatch selective detectors, or tunable detectors according to the practical scenarios.

1.4.3 The Presence of Interference

Apart from the noise, the received echoes often contain interference caused by civil broadcasting systems and electronic countermeasures. The signal modelling including the interference is more realistic. For radar systems, noise interference and coherent interference are two main types of interference. The noise interference is the high-power noise signal produced by the jammer. The noise interference can cover the target echo received by

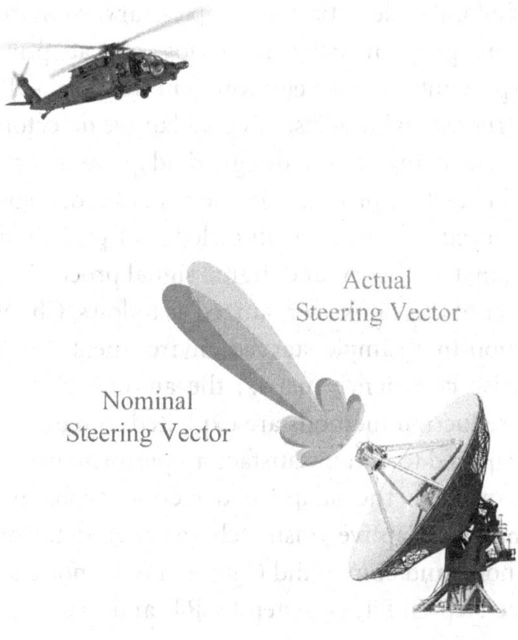

FIGURE 1.5 The schematic diagram of signal mismatch scenario.

the radar receiver such that it is difficult to obtain the target information from the received signals.

Coherent interference, which is also called false target interference, can deceive the radar by imitating the real target signal. In the adaptive detection algorithms, the coherent interference is also called the subspace interference since the coherent interference always lies in a known subspace. When coherent interference exists, the radar will receive not only the real target information but also the false target information like the false range, speed, and location. The radar cannot make a judgment since the information of false target and real target is mixed. Compared with noise interference, coherent interference is more concealed and less likely to be detected. Therefore, it is important to investigate how to achieve satisfactory detection performance in interference plus noise environments.

1.5 ORGANIZATION AND OUTLINE OF THE BOOK

The conventional adaptive detection algorithms suffer detection performance degradation when the ideal environment assumption is not satisfied. Following the growing demands on adaptive multichannel signals detection in non-ideal environments, this book provides a systematic presentation of adaptive detection in sample-starved environment, adaptive detection in signal mismatch scenarios, and adaptive detection in noise plus subspace interference environment. Readers will acquire skills in how to construct signal models, design adaptive detectors, and analyze the detection performance of the designed adaptive detectors when they deal with signal detection problems in the field of radar, sonar, and so on. Readers are required to have basic knowledge of probability theory, stochastic process, matrix theory, and digital signal processing.

The remainder of this book is organized as follows. Chapter 2 discusses adaptive detection in a sample-starved environment. The persymmetric structure of noise covariance matrix, the autoregressive property, and the dimension reduction methods are exploited to reduce the number of training data required to obtain satisfactory performance.

Chapter 3 considers the adaptive detection problem in the signal mismatch scenario. Adaptive mismatch selective detectors in partially homogeneous noise and compound Gaussian (CG) noise are designed by resorting to one-step GLRT, two-step GLRT, and maximum a posteriori (MAP) test.

Chapter 4 addresses the problem of detecting point-like targets in Gaussian noise when the signal mismatch occurs. An adaptive mismatch robust detector is designed according to the gradient test. The CFAR property analysis of the robust gradient detector is provided.

Chapter 5 introduces a kind of tunable detector, which is very flexible in controlling the detection performance for mismatched signals. The tunable detector with appropriate parameters can either be robust or selective. Moreover, the tunable detector can also provide higher PD than other detectors in the absence of signal mismatch.

Chapter 6 deals with adaptive signal detection problems in noise plus interference. We derive the adaptive detectors in signal-dependent interference based on Rao and Wald tests, adaptive detectors in signal-independent interference plus homogeneous and partially homogeneous clutter based on the GLRT and Wald test.

Chapter 7 is devoted to the analysis of the possible future development trend of adaptive detection methods in the three types of non-ideal scenarios. Meanwhile, two other further research tracks are also analyzed, namely, the adaptive detection algorithms based on the information geometry and the hybrid approaches that combine adaptive detection algorithms with machine learning algorithms.

REFERENCES

1. E. J. Kelly, "An Adaptive Detection Algorithm," *IEEE Transactions on Aerospace and Electronic Systems*, vol. 22, no. 2, pp. 115–127, 1986.
2. F. C. Robey, D. L. Fuhrman, E. J. Kelly, and R. Nitzberg, "A CFAR Adaptive Matched Filter Detector," *IEEE Transactions on Aerospace and Electronic Systems*, vol. 29, no. 1, pp. 208–216, 1992.
3. A. De Maio, "Rao Test for Adaptive Detection in Gaussian Interference with Unknown Covariance Matrix," *IEEE Transactions on Signal Processing*, vol. 55, no. 7, pp. 3577–3584, 2007.
4. A. De Maio, "A New Derivation of The Adaptive Matched Filter," *IEEE Signal Processing Letters*, vol. 11, no. 10, pp. 792–793, 2004.
5. H.-R. Park, J. Li, and H. Wang, "Polarization-Space-Time Domain Generalized Likelihood Ratio Detection of Radar Targets," *Signal Processing*, vol. 41, no. 2, pp. 153–164, 1995.
6. D. Pastina, P. Lombardo, and T. Bucciarelli, "Adaptive Polarimetric Target Detection with Coherent Radar Part I: Detection Against Gaussian Background," *IEEE Transactions on Aerospace and Electronic Systems*, vol. 37, no. 4, pp. 1194–1206, 2001.
7. A. De Maio and G. Ricci, "A Polarmetric Adaptive Matched Filter," *Signal Processing*, vol. 81, no.12, pp. 2583–2589, 2001.
8. X. Wang and H. V. Poor, "Robust Multiuser Detection in Non-Gaussian Channels," *IEEE Transactions on Signal Processing*, vol. 47, no. 2, pp. 289–305, 1999.
9. L. L. Scharf and B. Friedlander, "Matched Subspace Detectors," *IEEE Transactions on Signal Processing*, vol. 42, no. 8, pp. 2146–2157, 1994.
10. J.-J. Fuchs, "Multipath Time-Delay Detection and Estimation," *IEEE Transactions on Signal Processing*, vol. 47, no. 1, pp. 237–243, 1999.
11. E. Conte, A. De Maio, and G. Ricci, "GLRT-Based Adaptive Detection Algorithms for Range-Spread Targets," *IEEE Transactions on Signal Processing*, vol. 49, no. 7, pp. 1336–1348, 2001.
12. X. Shuai, L. Kong, and J. Yang, "Adaptive Detection for Distributed Targets in Gaussian Noise with Rao and Wald Tests," *Science China Information Sciences*, vol. 55, pp. 1290–1300, 2012.
13. O. Besson, L. L. Scharf, and S. Kraut, "Adaptive Detection of a Signal Known Only to Lie on a Line in a Known Subspace, When Primary and Secondary Data are Partially Homogeneous," *IEEE Transactions on Signal Processing*, vol. 54, no. 12, pp. 4698–4705, 2006.
14. E. Conte, A. De Maio, and C. Galdi, "CFAR Detection of Multidimensional Signals: An Invariant Approach," *IEEE Transactions on Signal Processing*, vol. 51, no. 1, pp. 142–151, 2003.
15. W. Liu, W. Xie, and Y. Wang, "Rao and Wald Tests for Distributed Targets Detection with Unknown Signal Steering," *IEEE Signal Processing Letters*, vol. 20, no. 11, pp. 1086–1089, 2013.

16. W. Liu, J. Liu, C. Hao, Y. Gao, and Y. Wang, "Multichannel Adaptive Signal Detection: Basic Theory and Literature Review," *Science China Information Sciences*, vol. 65, no. 2, pp. 1–41, 2022.
17. P. Lombardo and D. Pastina, "Adaptive Polarimetric Target Detection with Coherent Radar Part II: Detection Against Non-Gaussian Background," *IEEE Transactions on Aerospace and Electronic Systems*, vol. 37, no. 4, pp. 1207–1220, 2001.
18. J. Liu, Z.-J. Zhang, and Y. Yang, "Optimal Waveform Design for Generalized Likelihood Ratio and Adaptive Matched Filter Detectors Using a Diversely Polarized Antenna," *Signal Processing*, vol. 92, no. 4, pp. 1126–1131, 2012.
19. K. Gerlach and F. C. Lin, "Convergence Performance of Binary Adaptive Detectors," *IEEE Transactions on Aerospace and Electronic Systems*, vol. 31, no. 1, pp. 329–340, 1995.
20. S. Kraut and L. L. Scharf, "The CFAR Adaptive Subspace Detector is a Scale-Invariant GLRT," *IEEE Transactions on Signal Processing*, vol. 47, no. 9, pp. 2538–2541, 1999.
21. F. Pascal, Y. Chitour, J. Ovarlez, P. Forster, and P. Larzabal, "Covariance Structure Maximum-Likelihood Estimates in Compound Gaussian Noise: Existence and Algorithm Analysis," *IEEE Transactions on Signal Processing*, vol. 56, no. 1, pp. 34–48, 2008.
22. E. Conte, M. Lops, and G. Ricci, "Asymptotically Optimum Radar Detection in Compound-Gaussian Clutter," *IEEE Transactions on Aerospace and Electronic Systems*, vol. 31, no. 2, pp. 617–625, 1995.
23. A. De Maio and S. Iommelli, "Coincidence of the Rao Test, Wald Test, and GLRT in Partially Homogeneous Environment," *IEEE Signal Processing Letters*, vol. 15, pp. 385–388, 2008.
24. W. Liu, W. Xie, J. Liu, and Y. Wang, "Adaptive Double Subspace Signal Detection in Gaussian Background Part I: Homogeneous Environments," *IEEE Transactions on Signal Processing*, vol. 62, no. 9, pp. 2345–2357, 2014.
25. R. S. Raghavan, "Maximal Invariants and Performance of Some Invariant Hypothesis Tests for an Adaptive Detection Problem," *IEEE Transactions on Signal Processing*, vol. 61, no. 14, pp. 3607–3619, 2013.
26. S. Kraut and L. L. Scharf, "Adaptive Subspace Detectors," *IEEE Transactions on Signal Processing*, vol. 49, no. 1, pp. 1–16, 2001.
27. A. De Maio, "Invariance Theory for Adaptive Radar Detection in Heterogeneous Environment," *IEEE Signal Processing Letters*, vol. 26, no. 7, pp. 996–1000, 2019.
28. F. Gini, "Sub-Optimum Coherent Radar Detection in a Mixture of K-Distributed and Gaussian Clutter," *IEE Proceedings on Radar, Sonar and Navigation*, vol. 144, no. 1, pp. 39–48, 1997.
29. M. Rangaswamy, "Statistical Analysis of the Nonhomogeneity Detector for Non-Gaussian Interference Backgrounds," *IEEE Transactions on Signal Processing*, vol. 53, no. 6, pp. 2101–2111, 2005.
30. H. Wang and L. Cai, "On Adaptive Multiband Signal Detection with GLR Algorithm," *IEEE Transactions on Aerospace and Electronic Systems*, vol. 27, no. 2, pp. 225–233, 1991.

31. A. De Maio, G. Foglia, and E. Conte, "CFAR Behavior of Adaptive Detectors: An Experimental Analysis," *IEEE Transactions on Aerospace and Electronic Systems*, vol. 41, no. 1, pp. 233–251, 2005.
32. E. Conte, A. De Maio, and G. Ricci, "Recursive Estimation of the Covariance Matrix of a Compound-Gaussian Process and Its Application to Adaptive CFAR Detection," *IEEE Transactions on Signal Processing*, vol. 8, no. 50, pp. 1908–1915, 2002.
33. M. Sun, W. Liu, J. Liu, and C. Hao, "Complex Parameter Rao, Wald, Gradient, and Durbin Tests for Multichannel Signal Detection," *IEEE Transactions on Signal Processing*, vol. 70, pp. 117–131, 2022.

CHAPTER 2

Adaptive Detectors in Sample-Starved Environments

In Chapter 1, the block diagram and four commonly used design criteria for adaptive detection are introduced. To detect the signal in noise with an unknown covariance matrix, at least N training data are required to obtain a nonsingular sample covariance matrix, where N denotes the system freedom. In fact, at least $2N$ training data, free of signal components, are required to achieve satisfactory performance. With the improvement of manufacturing technology and the processing speed of hardware, the parameters related to the system freedom such as the number of channels in the radar system increase gradually, which makes the required training data increase sharply [1]. However, it is difficult to meet the requirement of sufficient training data since the practical scenario is complex and changeable. The adaptive detectors will suffer from detection performance degradation when the number of training data is limited.

The main methods to relax the requirement of the training data include dimension reduction methods, reduced-rank methods, parametric methods, prior knowledge-aided methods, and so on. The main idea of the dimension reduction method is to reduce the dimension of the data by projecting the high dimensional received data into the low dimensional space through the transformation matrix, which has nothing to do with the received data. Olivier Besson [2] applied a series of random

DOI: 10.1201/9781003477907-2

semi-unitary matrices to achieve dimension reduction and then derived the corresponding generalized likelihood ratio tests. The median of these generalized likelihood ratio tests is used as the final adaptive detector. Wang et al. [3] multiplied the received data by the conjugate transpose of the steering vector to eliminate interference outside the signal subspace and derived two new generalized likelihood ratio detectors based on the scalar data obtained by dimension reduction.

The reduced-rank method reduces the demand for training data by designing a transformation matrix related to echo data. Lin et al. [4] derived the reduced-rank conjugate gradient persymmetric AMF by projecting the persymmetric AMF into the Krylov subspace. Li et al. [5] first projected the data under test to the complement space of the clutter subspace, which was estimated by random matrix theory, by using a data-related projection matrix and then accumulated the energy of the whitened data to detect moving targets.

The autoregressive (AR) model, which represents the correlation between noise signals in the time domain, is an effective method to deal with the problem of limited training data. Hongbin Li et al. [6] conducted a series of research on the parametric methods based on multichannel AR model and proposed a parametric GLRT, parametric AMF, parametric Rao detector, and so on. Chengpeng Hao [7] modelled the noise as a multichannel AR Gaussian process and derived parametric detectors when the target energy leakages between adjacent range cells.

The principle of the adaptive Bayesian detectors is to derive adaptive detectors by assuming that the noise covariance matrix is a random variable and obeys a certain prior distribution or assuming that both the covariance matrix of data under test and the noise covariance matrix of the training data are random variables and that the relationship between the noise covariance matrices satisfies a certain probability distribution model [8]. Since the prior distribution exploits some prior knowledge of the environment in Bayesian detectors, the detection performance of the Bayesian detectors improves in the limited training cases. The exploitation of the covariance matrix structures such as persymmetric structure [9], Toeplitz structure [10], and Kronecker product structure [11] is also an effective way of improving detection performance or estimation accuracy of the covariance matrix.

This chapter discusses the design of adaptive detectors when the training data are insufficient. Section 2.1 and Section 2.2 introduce the design of the adaptive detectors for point-like targets and distributed targets in

AR noise. Section 2.3 gives the derivation of the detection scheme based on the orthogonal projection technique. Section 2.4 and Section 2.5 devote the design of adaptive persymmetric detectors for distributed targets and subspace signals in partially homogeneous environment (PHE).

2.1 ADAPTIVE POINT-LIKE TARGET DETECTION IN AUTOREGRESSIVE NOISE

In order to overcome the detection degradation for the conventional detectors in the sample-starved environments, an improved AMF is proposed by modelling the noise as an AR process with unknown parameters. The detector is derived by resorting to a two-step design procedure. First, the GLRT is derived under the assumption that the AR parameters are known. Then, the MLEs of the AR parameters, based on the training data, are substituted in place of the true AR parameters into the decision rule.

2.1.1 Detector Design

We assume that data are collected from a coherent train of N pulses. The detection problem at hand can be formulated by the following binary hypothesis testing

$$\begin{cases} H_0 : z_0 = \boldsymbol{n}_0 \\ H_1 : z_0 = \alpha \boldsymbol{p} + \boldsymbol{n}_0 \end{cases} \tag{2.1}$$

where $z_0 \in \mathbb{C}^{N \times 1}$ are the primary data under test, $\boldsymbol{p} = \left[1, e^{j\Omega}, \ldots, e^{j(N-1)\Omega}\right]^T$ denotes the steering vector, $\Omega = 2\pi f_d$, f_d denotes the normalized Doppler frequency, α is the unknown deterministic parameter accounting for the target reflectivity and the channel effects. Moreover, we assume that a set of training data $z_t = \boldsymbol{n}_t \in \mathbb{C}^{N \times 1}$, $t = 1, \ldots, K$ are available, $\boldsymbol{n}_t \in \mathbb{C}^{N \times 1}$, $t = 0, \ldots, K$ are independent zero-mean complex Gaussian vectors with unknown covariance matrix $\boldsymbol{R}$. In this book, the zero-mean complex Gaussian vector refers to the circularly symmetric complex Gaussian vector. Some properties of the circularly symmetric complex Gaussian vector are given in Appendix 2.A.

To improve detection performance, the noise signal $\boldsymbol{n}_t \in \mathbb{C}^{N \times 1}$, $t = 0, \ldots, K$, is modelled as an AR process of order M:

$$\boldsymbol{n}_t(l) = -\sum_{m=1}^{M} \boldsymbol{a}(m)\boldsymbol{n}_t(l-m) + \boldsymbol{w}_t(l) \qquad l = 1, \ldots, N \tag{2.2}$$

where $\boldsymbol{a} = \left[a(1),\ldots,a(M)\right]^T$ is the complex AR parameter vector, $w_t(l)$ is zero-mean complex white Gaussian noise with variance σ^2. Here, $\boldsymbol{a}$ and σ^2 are unknown constants.

It is shown that for $N >> M$ [12], the PDFs of z_0 under hypotheses H_0 and H_1 can be expressed as

$$f\left(z_0 \mid \boldsymbol{a},\sigma^2,H_0\right) = \frac{1}{(\pi\sigma^2)^{(N-M)}} \times \exp\left[-\frac{1}{\sigma^2}\left(\boldsymbol{u}_0 + \boldsymbol{Y}_0\boldsymbol{a}\right)^H\left(\boldsymbol{u}_0 + \boldsymbol{Y}_0\boldsymbol{a}\right)\right] \quad (2.3)$$

$$f\left(z_0 \mid \boldsymbol{a},\sigma^2,\alpha,H_1\right) = \frac{1}{(\pi\sigma^2)^{(N-M)}} \times\exp\left\{-\frac{1}{\sigma^2}\left[\boldsymbol{u}_0 + \boldsymbol{Y}_0\boldsymbol{a} - \alpha(\boldsymbol{q}+\boldsymbol{P}\boldsymbol{a})\right]^H\left[\boldsymbol{u}_0 + \boldsymbol{Y}_0\boldsymbol{a} - \alpha(\boldsymbol{q}+\boldsymbol{P}\boldsymbol{a})\right]\right\} \quad (2.4)$$

where $\boldsymbol{u}_t = \left[z_t(M+1),\ldots,z_t(N)\right]^T$, $t = 0,\ldots,K$, and $\boldsymbol{q} = \left[\boldsymbol{p}(M+1),\ldots,\boldsymbol{p}(N)\right]^T$ are two $\left(N-M\right)$-dimensional complex column vectors, $\boldsymbol{P} = \begin{pmatrix} \boldsymbol{p}(M) & \boldsymbol{p}(M-1) & \cdots & \boldsymbol{p}(1) \\ \boldsymbol{p}(M+1) & \boldsymbol{p}(M) & \cdots & \boldsymbol{p}(2) \\ \vdots & \vdots & \ddots & \vdots \\ \boldsymbol{p}(N-1) & \boldsymbol{p}(N-2) & \cdots & \boldsymbol{p}(N-M) \end{pmatrix}$ and

$\boldsymbol{Y}_t = \begin{pmatrix} z_t(M) & z_t(M-1) & \cdots & z_t(1) \\ z_t(M+1) & z_t(M) & \cdots & z_t(2) \\ \vdots & \vdots & \ddots & \vdots \\ z_t(N-1) & z_t(N-2) & \cdots & z_t(N-M) \end{pmatrix}$, $t = 0,\ldots,K$ are two $\left(N-M\right)\times M$ - dimensional matrices.

When $\boldsymbol{a}$ and σ^2 are known, the GLRT has the following form

$$\frac{\max_{\alpha} f\left(z_0 \mid \boldsymbol{a},\sigma^2,\alpha,H_1\right)}{f\left(z_0 \mid \boldsymbol{a},\sigma^2,H_0\right)} \underset{H_0}{\overset{H_1}{\gtrless}} \eta \quad (2.5)$$

where η denotes the detection threshold. From Equations 2.4 and 2.5, we can see that maximizing $f\left(z_0 \mid \boldsymbol{a},\sigma^2,\alpha,H_1\right)$ over α is equivalent to minimizing the expression $J(\alpha)=\left[\boldsymbol{u}_0+\boldsymbol{Y}_0\boldsymbol{a}-\alpha(\boldsymbol{q}+\boldsymbol{Pa})\right]^H\left[\boldsymbol{u}_0+\boldsymbol{Y}_0\boldsymbol{a}-\alpha(\boldsymbol{q}+\boldsymbol{Pa})\right]$ over α. We expand $J(\alpha)$ as follows

$$J(\alpha)=(\boldsymbol{q}+\boldsymbol{Pa})^H(\boldsymbol{q}+\boldsymbol{Pa})\times\left|\alpha-\frac{(\boldsymbol{q}+\boldsymbol{Pa})^H(\boldsymbol{u}_0+\boldsymbol{Y}_0\boldsymbol{a})}{(\boldsymbol{q}+\boldsymbol{Pa})^H(\boldsymbol{q}+\boldsymbol{Pa})}\right|^2 +(\boldsymbol{u}_0+\boldsymbol{Y}_0\boldsymbol{a})^H(\boldsymbol{u}_0+\boldsymbol{Y}_0\boldsymbol{a})-\frac{\left|(\boldsymbol{q}+\boldsymbol{Pa})^H(\boldsymbol{u}_0+\boldsymbol{Y}_0\boldsymbol{a})\right|^2}{(\boldsymbol{q}+\boldsymbol{Pa})^H(\boldsymbol{q}+\boldsymbol{Pa})} \tag{2.6}$$

The minimum is clearly attained when the positive factor containing α is made to vanish. Therefore, the MLE of α can be determined as

$$\hat{\alpha}=\frac{(\boldsymbol{q}+\boldsymbol{Pa})^H(\boldsymbol{u}_0+\boldsymbol{Y}_0\boldsymbol{a})}{(\boldsymbol{q}+\boldsymbol{Pa})^H(\boldsymbol{q}+\boldsymbol{Pa})} \tag{2.7}$$

Substituting the MLE of α into Equation 2.5, after some algebra, yields

$$\frac{\left|(\boldsymbol{q}+\boldsymbol{Pa})^H(\boldsymbol{u}_0+\boldsymbol{Y}_0\boldsymbol{a})\right|^2}{\sigma^2(\boldsymbol{q}+\boldsymbol{Pa})^H(\boldsymbol{q}+\boldsymbol{Pa})}\underset{H_0}{\overset{H_1}{\gtrless}}\log\eta \tag{2.8}$$

2.1.2 The Parameter Estimation of the Autoregressive Process

Next, the MLEs of $\boldsymbol{a}$ and σ^2 are obtained by using the training data alone. The joint conditional PDF of the training data can be written as

$$f\left(z_1,\ldots,z_K \mid \boldsymbol{a},\sigma^2\right)=\frac{1}{(\pi\sigma^2)^{(N-M)K}}\times\exp\left[-\frac{1}{\sigma^2}\sum_{t=1}^{K}(\boldsymbol{u}_t+\boldsymbol{Y}_t\boldsymbol{a})^H(\boldsymbol{u}_t+\boldsymbol{Y}_t\boldsymbol{a})\right] \tag{2.9}$$

Taking the logarithm of the joint conditional PDF and omitting the constants, we have

$$\ln f\left(z_1,\ldots,z_K \mid \boldsymbol{a},\sigma^2\right) = -(N-M)K\ln\sigma^2 - \frac{1}{\sigma^2}\sum_{t=1}^{K}\left(\boldsymbol{u}_t + \boldsymbol{Y}_t\boldsymbol{a}\right)^H\left(\boldsymbol{u}_t + \boldsymbol{Y}_t\boldsymbol{a}\right) \tag{2.10}$$

Taking the derivative of Equation 2.10 with respect to σ^2 and setting it to zero yields the MLE of σ^2

$$\hat{\sigma}^2 = \frac{1}{(N-M)K}\sum_{t=1}^{K}\left(\boldsymbol{u}_t + \boldsymbol{Y}_t\boldsymbol{a}\right)^H\left(\boldsymbol{u}_t + \boldsymbol{Y}_t\boldsymbol{a}\right) \tag{2.11}$$

Substituting Equation 2.11 into Equation 2.10 and omitting the constants yields

$$\ln f\left(z_1,\ldots,z_K \mid \boldsymbol{a}\right) = -(N-M)K\ln\left[\frac{1}{(N-M)K}\sum_{t=1}^{K}\left(\boldsymbol{u}_t + \boldsymbol{Y}_t\boldsymbol{a}\right)^H\left(\boldsymbol{u}_t + \boldsymbol{Y}_t\boldsymbol{a}\right)\right] \tag{2.12}$$

From Equation 2.12, we can see that the MLE of $\boldsymbol{a}$ can be obtained by minimizing the expression $\boldsymbol{Q}(\boldsymbol{a}) = \sum_{t=1}^{K}\left(\boldsymbol{u}_t + \boldsymbol{Y}_t\boldsymbol{a}\right)^H\left(\boldsymbol{u}_t + \boldsymbol{Y}_t\boldsymbol{a}\right)$ over $\boldsymbol{a}$. We expand the expression $\boldsymbol{Q}(\boldsymbol{a})$ as

$$\boldsymbol{Q}(\boldsymbol{a}) = \left(\boldsymbol{a}^H + \boldsymbol{S}_{Yu}^H\boldsymbol{S}_{YY}^{-1}\right)\boldsymbol{S}_{YY}\left(\boldsymbol{a}^H + \boldsymbol{S}_{Yu}^H\boldsymbol{S}_{YY}^{-1}\right)^H - \boldsymbol{S}_{Yu}^H\boldsymbol{S}_{YY}^{-1}\boldsymbol{S}_{Yu} + \boldsymbol{S}_{uu} \tag{2.13}$$

where $\boldsymbol{S}_{uu} = \sum_{t=1}^{K}\boldsymbol{u}_t^H\boldsymbol{u}_t$, $\boldsymbol{S}_{Yu} = \sum_{t=1}^{K}\boldsymbol{Y}_t^H\boldsymbol{u}_t$, and $\boldsymbol{S}_{YY} = \sum_{t=1}^{K}\boldsymbol{Y}_t^H\boldsymbol{Y}_t$. Since $\boldsymbol{S}_{YY}$ is nonnegative definite, it follows that

$$\hat{\boldsymbol{a}} = -\left(\sum_{t=1}^{K}\boldsymbol{Y}_t^H\boldsymbol{Y}_t\right)^{-1}\left(\sum_{t=1}^{K}\boldsymbol{Y}_t^H\boldsymbol{u}_t\right) \tag{2.14}$$

Finally, after some calculation, we can get the proposed model-based adaptive matched filter (MAMF) detector

$$\frac{K(N-M)\left|(\boldsymbol{q}+\boldsymbol{P}\hat{\boldsymbol{a}})^H(\boldsymbol{u}_0+\boldsymbol{Y}_0\hat{\boldsymbol{a}})\right|^2}{\left[\sum_{t=1}^{K}(\boldsymbol{u}_t+\boldsymbol{Y}_t\hat{\boldsymbol{a}})^H(\boldsymbol{u}_t+\boldsymbol{Y}\hat{\boldsymbol{a}})\right]\cdot(\boldsymbol{q}+\boldsymbol{P}\hat{\boldsymbol{a}})^H(\boldsymbol{q}+\boldsymbol{P}\hat{\boldsymbol{a}})} \underset{H_0}{\overset{H_1}{\gtrless}} \eta_{MAMF} \tag{2.15}$$

where η_{MAMF} denotes the detection threshold.

2.1.3 Numerical Examples

A simulation study is carried out to investigate the proposed MAMF detector. The SNR is defined as $\text{SNR} = |\alpha|^2 \boldsymbol{p}^H \boldsymbol{R}^{-1} \boldsymbol{p}$, where $\boldsymbol{R}$ is the covariance matrix of the noise and can be computed using $\boldsymbol{a}$ and σ^2. To decrease the computation load, we set $P_{fa} = 10^{-2}$. The PD and the threshold are determined by resorting to 10^4 and 10^5 independent Monte Carlo trials, respectively.

In Figure 2.1, we compare the performance of the MAMF with the conventional AMF and GLRT. From Figure 2.1, we can see that with the increase of K, the performance of the three detectors becomes better and better. In Figure 2.1(a), the conventional detectors are not plotted since in this case the sample covariance matrix used in the conventional detectors is singular. From Figure 2.1(b) and Figure 2.1(c), we can see that when the number of training data is small, the proposed MAMF achieves a significant performance improvement over the conventional detectors. This illustrates that exploiting the property of the noise can improve the detection performance in the sample-starved environment. The curves in Figure 2.1(d) show that both the MAMF and the conventional detectors can work well when the training data are sufficient.

Figure 2.2 shows the detection performance versus the SNR for various N. For comparison purposes, we also show the performance of the optimal matched filter (MF). The MF cannot be used in practice but offers a baseline for comparison. From Figure 2.2, it can be seen that with the increase of N, the performance of the proposed MAMF becomes better and better. For $K = 1, N = 40, P_d = 0.9$, the performance loss of the proposed MAMF with respect to the MF is less than 1dB as shown in Figure 2.2 (a). In Figure 2.2(b), the performance gap between the MAMF and the MF is smaller. Hence, when N is moderate, the detection performance of the proposed MAMF is close to the MF even in the sample-starved scenarios.

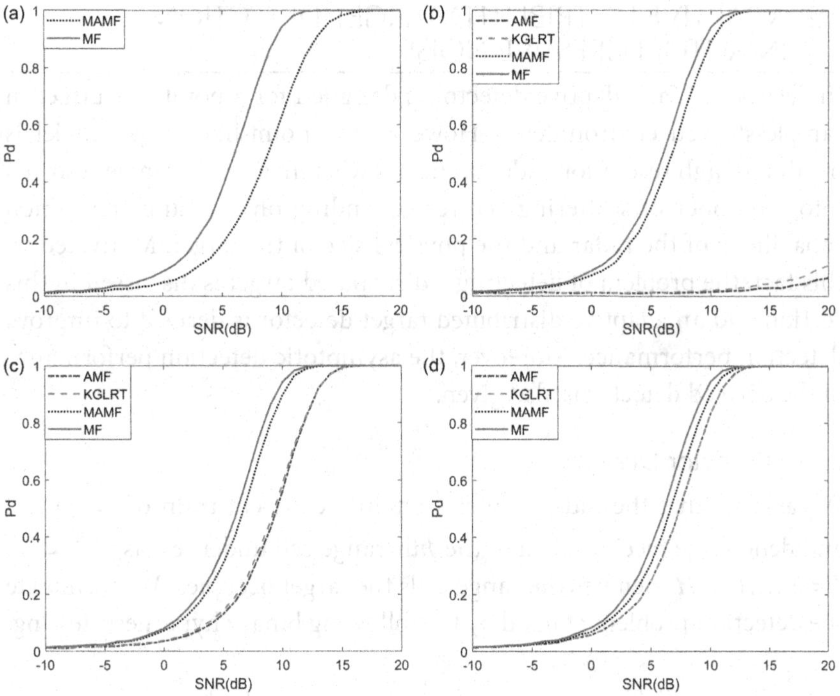

FIGURE 2.1 PD versus SNR for $N = 20$. (a) $K = 1$; (b) $K = 20$; (c) $K = 40$; (d) $K = 80$.

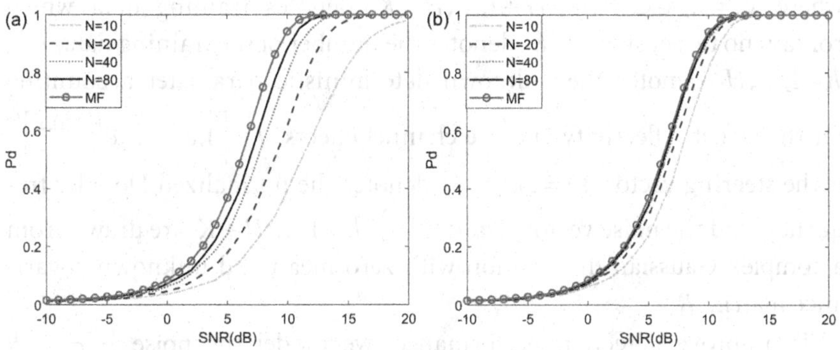

FIGURE 2.2 PD versus SNR for various N. (a) $K = 1$; (b) $K = 20$.

2.2 ADAPTIVE DISTRIBUTED TARGET DETECTION IN AUTOREGRESSIVE NOISE

In Section 2.1, an adaptive detector is designed for a point-like target in sample-starved environments. However, the point-like target model is invalid in high resolution radar scenarios wherein a target can be resolved into a number of scattering centres depending on the range resolution capabilities of the radar and the physical size of the target. Motivated by this fact, the problem of detecting a distributed target is discussed in this section and an adaptive distributed target detector is derived to improve detection performance. Moreover, the asymptotic detection performance of the derived detector is also given.

2.2.1 Detector Design

We assume that the radar echoes contain a coherent train of N pulses and denote the received data in the hth range cell under test as $\boldsymbol{y}_h \in \mathbb{C}^{N\times 1}$, $h=1,\ldots,H$, H denotes the range cells the target occupies. We formulate the detection problem at hand by the following binary hypothesis testing

$$\begin{cases} H_0 : \boldsymbol{y}_h = \boldsymbol{n}_h, & h=1,\ldots,H+K, \\ H_1 : \begin{cases} \boldsymbol{y}_h = \alpha_h \boldsymbol{t} + \boldsymbol{n}_h, & h=1,\ldots,H, \\ \boldsymbol{y}_h = \boldsymbol{n}_h, & h=H+1,\ldots H+K, \end{cases} \end{cases} \tag{2.16}$$

where $\boldsymbol{y}_h = \boldsymbol{n}_h \in \mathbb{C}^{N\times 1}, h=H+1,\ldots,H+K$ denotes training data which contain no target signals, K denotes the number of the training data, α_h, $h=1,\ldots,H$ denotes the unknown deterministic parameter accounting for the target reflectivity and the channel effects, $\boldsymbol{t} = \left[1, e^{j\Omega}, \ldots, e^{j(N-1)\Omega}\right]^T$ is the steering vector, $\Omega = 2\pi f_d$, f_d denotes the normalized Doppler frequency, and the noise vectors $\left\{\boldsymbol{n}_h \in \mathbb{C}^{N\times 1}\right\}$, $h=1,\ldots,H+K$ are drawn from a complex Gaussian distribution with zero mean and unknown covariance matrix $\boldsymbol{R}$.

To improve detection performance, we model the noise $\boldsymbol{n}_h \in \mathbb{C}^{N\times 1}$, $h=1,\ldots,H+K$ as an AR process

$$\boldsymbol{n}_h(l) = -\sum_{m=1}^{M} \boldsymbol{a}(m)\boldsymbol{n}_h(l-m) + w_h(l) \quad l=1,\ldots,N \tag{2.17}$$

where $\boldsymbol{a}=\left[\boldsymbol{a}(1),\ldots,\boldsymbol{a}(M)\right]^T$ is the complex AR parameter vector, M is the order of the AR process, $w_h(l)$ is a sequence of complex white Gaussian noise with zero-mean and variance σ^2, $\boldsymbol{a}$ and σ^2 are unknown fix constants.

We set $\boldsymbol{\theta}_r=[\alpha_{1,R},\alpha_{1,I},\ldots,\alpha_{H,R},\alpha_{H,I}]^T\in\mathbb{R}^{2H\times1}$, $\boldsymbol{\theta}_s=[\boldsymbol{a}_R^T,\boldsymbol{a}_I^T,\sigma^2]^T\in\mathbb{R}^{(2M+1)\times1}$, $\boldsymbol{\theta}=\left[\boldsymbol{\theta}_r^T,\boldsymbol{\theta}_s^T\right]^T\in\mathbb{R}^{(2M+2H+1)\times1}$, $\boldsymbol{a}_R=\mathrm{vec}\left(\mathrm{Re}\{\boldsymbol{a}\}\right)$, $\boldsymbol{a}_I=\mathrm{vec}\left(\mathrm{Im}\{\boldsymbol{a}\}\right)$, $\alpha_{h,R}$ and $\alpha_{h,I}$ denote the real part and imaginary part of α_h $h=1,\ldots,H$, respectively.

To solve the detection problem in Equation 2.16, we resort to the Rao test based on the primary and training data, which is tantamount to the following decision rule [13]

$$\left.\frac{\partial\ln f(\boldsymbol{Y}\mid\boldsymbol{\theta})}{\partial\boldsymbol{\theta}_r}\right|^T_{\boldsymbol{\theta}=\hat{\boldsymbol{\theta}}_0}\left[\boldsymbol{J}^{-1}\left(\hat{\boldsymbol{\theta}}_0\right)\right]_{\boldsymbol{\theta}_r,\boldsymbol{\theta}_r}\times\left.\frac{\partial\ln f(\boldsymbol{Y}\mid\boldsymbol{\theta})}{\partial\boldsymbol{\theta}_r}\right|_{\boldsymbol{\theta}=\hat{\boldsymbol{\theta}}_0}\underset{H_0}{\overset{H_1}{\gtrless}}\eta_{MRao}\tag{2.18}$$

where $\boldsymbol{Y}=\left[\boldsymbol{y}_1,\boldsymbol{y}_2,\ldots,\boldsymbol{y}_{H+K}\right]$ denote the primary data and the training data, η_{MRao} is the detection threshold, $\partial/\partial\boldsymbol{\theta}_r=\left[\partial/\partial\alpha_{1,R},\partial/\partial\alpha_{1,I},\ldots,\partial/\partial\alpha_{H,R},\partial/\partial\alpha_{H,I}\right]^T$ denotes the gradient with respect to $\boldsymbol{\theta}_r$, $\hat{\boldsymbol{\theta}}_0=\left[\hat{\boldsymbol{\theta}}_{r,0}^T,\hat{\boldsymbol{\theta}}_{s,0}^T\right]^T$ is the MLE of $\boldsymbol{\theta}$ under H_0, $\boldsymbol{J}(\boldsymbol{\theta})=\boldsymbol{J}(\boldsymbol{\theta}_r,\boldsymbol{\theta}_s)$ denotes the FIM, and $f(\cdot)$ is the PDF for the data.

As shown in [12, 14, 15], the joint probability density function (PDF) for the primary and training data under H_i can be expressed as

$$f\left(\boldsymbol{Y}\mid\boldsymbol{a},\sigma^2,H_i\right)=\frac{1}{(\pi\sigma^2)^{(N-M)(H+K)}}\exp\left\{-\frac{1}{\sigma^2}\left[\sum_{h=H+1}^{H+K}\left(\boldsymbol{u}_h+\boldsymbol{X}_h\boldsymbol{a}\right)^H\left(\boldsymbol{u}_h+\boldsymbol{X}_h\boldsymbol{a}\right)\right.\right.$$
$$\left.\left.+\sum_{h=1}^{H}\left(\boldsymbol{u}_h+\boldsymbol{X}_h\boldsymbol{a}-i\alpha(\boldsymbol{q}+\boldsymbol{P}\boldsymbol{a})\right)^H\left(\boldsymbol{u}_h+\boldsymbol{X}_h\boldsymbol{a}-i\alpha(\boldsymbol{q}+\boldsymbol{P}\boldsymbol{a})\right)\right]\right\}\tag{2.19}$$

where $i=0,1$, $\boldsymbol{P}=\begin{pmatrix} t(M) & t(M-1) & \cdots & t(1) \\ t(M+1) & t(M) & \cdots & t(2) \\ \vdots & \vdots & \ddots & \vdots \\ t(N-1) & t(N-2) & \cdots & t(N-M) \end{pmatrix}$,

$$\boldsymbol{X}_h=\begin{pmatrix} y_h(M) & y_h(M-1) & \cdots & y_h(1) \\ y_h(M+1) & y_h(M) & \cdots & y_h(2) \\ \vdots & \vdots & \ddots & \vdots \\ y_h(N-1) & y_h(N-2) & \cdots & y_h(N-M) \end{pmatrix},$$

$\boldsymbol{q}=\left[t(M+1),\ldots,t(N)\right]^T$, $\boldsymbol{u}_h=\left[y_h(M+1),\ldots,y_h(N)\right]^T$,

$h=1,\ldots,H+K$.

2.2.2 Fisher Information Matrix

To obtain the Rao test, the FIM is calculated in this part.

We partition the FIM as $\boldsymbol{J}(\boldsymbol{\theta})=\begin{bmatrix} \boldsymbol{J}_{\theta_r,\theta_r}(\boldsymbol{\theta}) & \boldsymbol{J}_{\theta_r,\theta_s}(\boldsymbol{\theta}) \\ \boldsymbol{J}_{\theta_s,\theta_r}(\boldsymbol{\theta}) & \boldsymbol{J}_{\theta_s,\theta_s}(\boldsymbol{\theta}) \end{bmatrix}$,

where $\boldsymbol{J}_{\theta_r,\theta_r}(\boldsymbol{\theta})=-E\left[\dfrac{\partial^2 \ln f(\boldsymbol{Y}|\boldsymbol{\theta})}{\partial\boldsymbol{\theta}_r\partial\boldsymbol{\theta}_r^T}\right]$, $\boldsymbol{J}_{\theta_r,\theta_s}(\boldsymbol{\theta})=-E\left[\dfrac{\partial^2 \ln f(\boldsymbol{Y}|\boldsymbol{\theta})}{\partial\boldsymbol{\theta}_r\partial\boldsymbol{\theta}_s^T}\right]$,

$\boldsymbol{J}_{\theta_s,\theta_r}(\boldsymbol{\theta})=-E\left[\dfrac{\partial^2 \ln f(\boldsymbol{Y}|\boldsymbol{\theta})}{\partial\boldsymbol{\theta}_s\partial\boldsymbol{\theta}_r^T}\right]$, $\boldsymbol{J}_{\theta_s,\theta_s}(\boldsymbol{\theta})=-E\left[\dfrac{\partial^2 \ln f(\boldsymbol{Y}|\boldsymbol{\theta})}{\partial\boldsymbol{\theta}_s\partial\boldsymbol{\theta}_s^T}\right]$.

We take the first partial derivative of the logarithm of PDF $\ln f(\boldsymbol{Y}|\boldsymbol{\theta})$ with respect to $\boldsymbol{\theta}_r$ first. The result is given as

$$\frac{\partial \ln f(\boldsymbol{Y}|\boldsymbol{\theta})}{\partial\boldsymbol{\theta}_r}=\left[\frac{\partial \ln f(\boldsymbol{Y}|\boldsymbol{\theta})}{\partial\alpha_{1,R}},\frac{\partial \ln f(\boldsymbol{Y}|\boldsymbol{\theta})}{\partial\alpha_{1,I}},\ldots,\frac{\partial \ln f(\boldsymbol{Y}|\boldsymbol{\theta})}{\partial\alpha_{H,R}},\frac{\partial \ln f(\boldsymbol{Y}|\boldsymbol{\theta})}{\partial\alpha_{H,I}}\right]^T \tag{2.20}$$

where

$$\frac{\partial \ln f(\boldsymbol{Y}|\boldsymbol{\theta})}{\partial\alpha_{h,R}}=2\,\mathrm{Re}\left\{(\boldsymbol{q}+\boldsymbol{Pa})^H\cdot\left[\boldsymbol{u}_h+\boldsymbol{X}_h\boldsymbol{a}-\alpha_h(\boldsymbol{q}+\boldsymbol{Pa})\right]/\sigma^2\right\} \tag{2.21}$$

$$\frac{\partial \ln f(\mathbf{Y}|\boldsymbol{\theta})}{\partial \alpha_{h,I}} = 2\operatorname{Im}\left\{(\mathbf{q}+\mathbf{Pa})^H \cdot \left[\mathbf{u}_h + \mathbf{X}_h\mathbf{a} - \alpha_h(\mathbf{q}+\mathbf{Pa})\right]/\sigma^2\right\} \quad (2.22)$$

$\left[\mathbf{J}^{-1}(\boldsymbol{\theta})\right]_{\boldsymbol{\theta}_r,\boldsymbol{\theta}_r}$ in Equation 2.18 is the $(\boldsymbol{\theta}_r,\boldsymbol{\theta}_r)$-part of the inversion of $\mathbf{J}(\boldsymbol{\theta})$, namely,

$$\left[\mathbf{J}^{-1}(\boldsymbol{\theta})\right]_{\boldsymbol{\theta}_r,\boldsymbol{\theta}_r} = \left(\mathbf{J}_{\boldsymbol{\theta}_r,\boldsymbol{\theta}_r}(\boldsymbol{\theta}) - \mathbf{J}_{\boldsymbol{\theta}_r,\boldsymbol{\theta}_s}(\boldsymbol{\theta})\mathbf{J}^{-1}_{\boldsymbol{\theta}_s,\boldsymbol{\theta}_s}(\boldsymbol{\theta})\mathbf{J}_{\boldsymbol{\theta}_s,\boldsymbol{\theta}_r}(\boldsymbol{\theta})\right)^{-1} \quad (2.23)$$

In order to obtain $\left[\mathbf{J}^{-1}(\boldsymbol{\theta})\right]_{\boldsymbol{\theta}_r,\boldsymbol{\theta}_r}$, we take the partial derivative of $\partial \ln f(\mathbf{Y}|\boldsymbol{\theta})/\partial\boldsymbol{\theta}_r$ with respect to $\boldsymbol{\theta}_r$ and $\boldsymbol{\theta}_s$, respectively. After some algebra, we have

$$\frac{\partial^2 \ln f(\mathbf{Y}|\boldsymbol{\theta})}{\partial \alpha_{h1,R}\partial \alpha_{h2,I}} = \frac{\partial^2 \ln f(\mathbf{Y}|\boldsymbol{\theta})}{\partial \alpha_{h1,I}\partial \alpha_{h2,R}} = 0 \quad (2.24)$$

$$\frac{\partial^2 \ln f(\mathbf{Y}|\boldsymbol{\theta})}{\partial \alpha_{h1,R}\partial \alpha_{h2,R}} = \frac{\partial^2 \ln f(\mathbf{Y}|\boldsymbol{\theta})}{\partial \alpha_{h1,I}\partial \alpha_{h2,I}} = -2\delta(h1-h2)(\mathbf{q}+\mathbf{Pa})^H(\mathbf{q}+\mathbf{Pa})/\sigma^2 \quad (2.25)$$

$$\frac{\partial^2 \ln f(\mathbf{Y}|\boldsymbol{\theta})}{\partial \alpha_{h,R}\partial \sigma^2} = -2\operatorname{Re}\left\{(\mathbf{q}+\mathbf{Pa})^H \cdot \left[\mathbf{u}_h + \mathbf{X}_h\mathbf{a} - \alpha_h(\mathbf{q}+\mathbf{Pa})\right]/\sigma^4\right\} \quad (2.26)$$

$$\frac{\partial^2 \ln f(\mathbf{Y}|\boldsymbol{\theta})}{\partial \alpha_{h,R}\partial a_{R_i}} = 2\operatorname{Re}\left\{\left[\left(\mathbf{u}_h + \mathbf{X}_h\mathbf{a} - \alpha_h(\mathbf{q}+\mathbf{Pa})\right)^H \mathbf{P}\frac{\partial \mathbf{a}}{\partial a_{R_i}} + (\mathbf{q}+\mathbf{Pa})^H(\mathbf{X}_h - \alpha_h\mathbf{P})\frac{\partial \mathbf{a}}{\partial a_{R_i}}\right]/\sigma^2\right\} \quad (2.27)$$

where $h = 1,\ldots,H$, a_{R_i} is the ith element of $\mathbf{a}_R$. By taking the mathematical expectation of Equations 2.26 and 2.27, we can get $E\left[\mathbf{u}_h + \mathbf{X}_h\mathbf{a} - \alpha_h(\mathbf{q}+\mathbf{Pa})\right] = E\left[\mathbf{w}_h\right] = 0$,

$E\left[\boldsymbol{u}_h - \alpha_h \boldsymbol{q}\right] = 0$, $E\left[\boldsymbol{X}_h - \alpha_h \boldsymbol{P}\right] = 0$, $E\left\{\frac{\partial^2 \ln f(\boldsymbol{Y}|\boldsymbol{\theta})}{\partial \alpha_{h,R} \partial \sigma^2}\right\} = 0$, $E\left\{\frac{\partial^2 \ln f(\boldsymbol{Y}|\boldsymbol{\theta})}{\partial \alpha_{h,R} \partial \boldsymbol{a}_R}\right\} = 0$, $h = 1,\ldots,H$. Similarly, after some calculation, it can be shown that $E\left\{\frac{\partial^2 \ln f(\boldsymbol{Y}|\boldsymbol{\theta})}{\partial \alpha_{h,I} \partial \sigma^2}\right\} = 0$, $E\left\{\frac{\partial^2 \ln f(\boldsymbol{Y}|\boldsymbol{\theta})}{\partial \alpha_{h,I} \partial \boldsymbol{a}_R}\right\} = 0$, $E\left\{\frac{\partial^2 \ln f(\boldsymbol{Y}|\boldsymbol{\theta})}{\partial \alpha_{h,R} \partial \boldsymbol{a}_I}\right\} = 0$, $E\left\{\frac{\partial^2 \ln f(\boldsymbol{Y}|\boldsymbol{\theta})}{\partial \alpha_{h,I} \partial \boldsymbol{a}_I}\right\} = 0$. Substituting the above results into the FIM, we have $\boldsymbol{J}_{\theta_r,\theta_s}(\boldsymbol{\theta}) = \boldsymbol{0}$, $\boldsymbol{J}_{\theta_s,\theta_r}(\boldsymbol{\theta}) = \boldsymbol{0}$, $\boldsymbol{J}_{\theta_r,\theta_r}(\boldsymbol{\theta}) = \left(2(\boldsymbol{q}+\boldsymbol{P}\boldsymbol{a})^H(\boldsymbol{q}+\boldsymbol{P}\boldsymbol{a})/\sigma^2\right)\mathbf{I}_{2H\times 2H}$.

According to Equation 2.23, the term $\left[\boldsymbol{J}^{-1}(\boldsymbol{\theta})\right]_{\theta_r,\theta_r}$ can be calculated as

$$\left[\boldsymbol{J}^{-1}(\boldsymbol{\theta})\right]_{\theta_r,\theta_r} = \boldsymbol{J}^{-1}_{\theta_r,\theta_r}(\boldsymbol{\theta}) = \frac{1}{2}\left(\frac{1}{\sigma^2}(\boldsymbol{q}+\boldsymbol{P}\boldsymbol{a})^H(\boldsymbol{q}+\boldsymbol{P}\boldsymbol{a})\right)^{-1}\mathbf{I}_{2H\times 2H} \tag{2.28}$$

2.2.3 The Maximum Likelihood Estimates of the Unknown Parameters

In this part, the MLEs of σ^2 and $\boldsymbol{a}$ under hypothesis H_0 are calculated. By taking the derivative of the logarithm of the joint PDF $\ln f(\boldsymbol{Y}|\boldsymbol{\theta},H_0)$ with respect to σ^2 and equating the result to zero, we can obtain the MLE of σ^2 given $\boldsymbol{a}$

$$\hat{\sigma}^2 = \frac{1}{(N-M)(H+K)}\left[\sum_{h=1}^{H+K}(\boldsymbol{u}_h + \boldsymbol{X}_h\boldsymbol{a})^H(\boldsymbol{u}_h + \boldsymbol{X}_h\boldsymbol{a})\right] \tag{2.29}$$

Substituting $\hat{\sigma}^2$ into $f(\boldsymbol{Y}|\boldsymbol{a},H_0)$ yields

$$\max_{\sigma^2} f(\boldsymbol{Y}|\boldsymbol{a},H_0)$$

$$= \left(\pi e \frac{1}{(N-M)(H+K)}\left[\sum_{h=1}^{H+K}(\boldsymbol{u}_h + \boldsymbol{X}_h\boldsymbol{a})^H(\boldsymbol{u}_h + \boldsymbol{X}_h\boldsymbol{a})\right]\right)^{-(N-M)(H+K)} \tag{2.30}$$

From Equation 2.30, the MLE of $\boldsymbol{a}$ under H_0 can be obtained by minimizing $\boldsymbol{T} = \sum_{h=1}^{H+K} (\boldsymbol{u}_h + \boldsymbol{X}_h \boldsymbol{a})^H (\boldsymbol{u}_h + \boldsymbol{X}_h \boldsymbol{a})$ with respect to $\boldsymbol{a}$. We expand the expression $\boldsymbol{T}$ as follows

$$\begin{aligned} \boldsymbol{T} &= \boldsymbol{S}_{uu} + \boldsymbol{S}_{Xu}^H \boldsymbol{a} + \boldsymbol{a}^H \boldsymbol{S}_{Xu} + \boldsymbol{a}^H \boldsymbol{S}_{XX} \boldsymbol{a} \\ &= \left(\boldsymbol{a}^H + \boldsymbol{S}_{Xu}^H \boldsymbol{S}_{XX}^{-1}\right) \boldsymbol{S}_{XX} \left(\boldsymbol{a}^H + \boldsymbol{S}_{Xu}^H \boldsymbol{S}_{XX}^{-1}\right)^H - \boldsymbol{S}_{Xu}^H \boldsymbol{S}_{XX}^{-1} \boldsymbol{S}_{Xu} + \boldsymbol{S}_{uu} \end{aligned} \tag{2.31}$$

where $\boldsymbol{S}_{uu} = \sum_{h=1}^{H+K} \boldsymbol{u}_h^H \boldsymbol{u}_h$, $\boldsymbol{S}_{Xu} = \sum_{h=1}^{H+K} \boldsymbol{X}_h^H \boldsymbol{u}_h$, $\boldsymbol{S}_{XX} = \sum_{h=1}^{H+K} \boldsymbol{X}_h^H \boldsymbol{X}_h$. Since $\boldsymbol{S}_{XX}$ is a nonnegative definite matrix and the second and third terms in Equation 2.31 do not depend on $\boldsymbol{a}$, it follows that [16]

$$\hat{\boldsymbol{a}} = -\boldsymbol{S}_{XX}^{-1} \boldsymbol{S}_{Xu} = -\left(\sum_{h=1}^{H+K} \boldsymbol{X}_h^H \boldsymbol{X}_h\right)^{-1} \left(\sum_{h=1}^{H+K} \boldsymbol{X}_h^H \boldsymbol{u}_h\right) \tag{2.32}$$

Finally, by substituting Equations 2.20–2.22 and Equations 2.28–2.32 into the decision rule Equation 2.18 and rearranging the expression, we can obtain the model-based Rao (MRao) detector

$$\frac{2(N-M)(H+K)\sum_{h=1}^{H} \left|(\boldsymbol{q}+\boldsymbol{P}\hat{\boldsymbol{a}})^H (\boldsymbol{u}_h + \boldsymbol{X}_h \hat{\boldsymbol{a}})\right|^2}{\left[\sum_{h=1}^{H+K} (\boldsymbol{u}_h + \boldsymbol{X}_h \hat{\boldsymbol{a}})^H (\boldsymbol{u}_h + \boldsymbol{X}_h \hat{\boldsymbol{a}})\right] \cdot (\boldsymbol{q}+\boldsymbol{P}\hat{\boldsymbol{a}})^H (\boldsymbol{q}+\boldsymbol{P}\hat{\boldsymbol{a}})} \underset{H_0}{\overset{H_1}{\gtrless}} \eta_{MRao} \tag{2.33}$$

2.2.4 Asymptotic Performance Analysis

We analyze the asymptotic performance of the proposed MRao detector in this section. The asymptotic expressions for the PFA and PD are derived. For large data records (i.e. N approaches infinity), the Rao detector has the same asymptotic performance as the GLRT [13, 14]

$$T_{MRao} \overset{a}{\sim} \begin{cases} \chi_{2H}^2, & H_0 \\ \chi_{2H}'^2(\lambda), & H_1 \end{cases} \tag{2.34}$$

where χ_{2H}^2 denotes the central Chi-squared distribution with $2H$ degrees of freedom and $\chi_{2H}^{\prime 2}(\lambda)$ the noncentral Chi-squared distribution with $2H$ degrees of freedom and noncentrality parameter λ

$$\lambda=(\boldsymbol{\theta}_{r1}-\boldsymbol{\theta}_{r0})^T\left[\boldsymbol{J}_{\theta_r\theta_r}(\boldsymbol{\theta}_{r0},\boldsymbol{\theta}_s)-\boldsymbol{J}_{\theta_r\theta_s}(\boldsymbol{\theta}_{r0},\boldsymbol{\theta}_s)\boldsymbol{J}_{\theta_s\theta_s}^{-1}(\boldsymbol{\theta}_{r0},\boldsymbol{\theta}_s)\boldsymbol{J}_{\theta_s\theta_r}(\boldsymbol{\theta}_{r0},\boldsymbol{\theta}_s)\right](\boldsymbol{\theta}_{r1}-\boldsymbol{\theta}_{r0}) \tag{2.35}$$

where $\boldsymbol{\theta}_{r1}$ and $\boldsymbol{\theta}_{r0}$ denote $\boldsymbol{\theta}_r$ under H_1 and H_0, and $(\boldsymbol{\theta}_{r1}-\boldsymbol{\theta}_{r0})=[\alpha_{1,R},\alpha_{1,I},\ldots,\alpha_{H,R},\alpha_{H,I}]^T$. Substituting the detailed expressions of the FIM given in Section 2.2.2 into (2.35), we can obtain the noncentrality parameter $\lambda=2\sum_{h=1}^{H}|\alpha_h|^2(\boldsymbol{q}+\boldsymbol{Pa})^H(\boldsymbol{q}+\boldsymbol{Pa})/\sigma^2$.

The asymptotic PFA can be given as

$$P_{fa}=\int_{\eta_{MRao}}^{\infty}p_{\chi_{2H}^2}(x)dx=\int_{\eta_{MRao}}^{\infty}\frac{1}{2^H\Gamma(H)}x^{H-1}\exp\left(-\tfrac{1}{2}x\right)dx \tag{2.36}$$

The asymptotic PD is

$$P_d=\int_{\eta_{MRao}}^{\infty}p_{\chi_{2H}^{\prime 2}(\lambda)}(x)dx=\int_{\eta_{MRao}}^{\infty}\frac{1}{2}\left(\frac{x}{\lambda}\right)^{(H-1)/2}\exp\left[-\tfrac{1}{2}(x+\lambda)\right]I_{H-1}\left(\sqrt{\lambda x}\right)dx \tag{2.37}$$

where $I_r(\cdot)$ is the modified Bessel function of the first kind and order r, defined as $I_r(u)=\frac{\left(\frac{1}{2}u\right)^r}{\sqrt{\pi}\Gamma\left(r+\frac{1}{2}\right)}\int_0^{\pi}\exp(u\cos\theta)\sin^{2r}\theta d\theta=\sum_{k=0}^{\infty}\frac{\left(\frac{1}{2}u\right)^{2k+r}}{k!\Gamma(r+k+1)}$, where $\Gamma(\cdot)$ denotes the gamma function. From Equation 2.36, it can be seen that the asymptotic PFA is independent of the covariance matrix. This indicates that the proposed MRao detector is asymptotically CFAR with respect to the covariance matrix.

2.2.5 Numerical Examples

Simulated data are resorted to assess the detection performance of the proposed MRao detector. For the simulated results, 10^4 independent Monte Carlo trials are conducted to evaluate the thresholds and 10^5 independent trials to compute the PD. The noise $\boldsymbol{n}_h$ are generated from the AR process of order two (unless otherwise specified) with given parameters $\boldsymbol{a}$ and σ^2. These parameters should be set to ensure that the AR process is stable. The true covariance matrix of the noise $\boldsymbol{R}$ can be computed using parameters $\boldsymbol{a}$ and σ^2 [17]. The SNR is defined as $\text{SNR} = \sum_{h=1}^{H} |\alpha_h|^2 \boldsymbol{t}^H \boldsymbol{R}^{-1} \boldsymbol{t} \big/ H$.

For comparison, the detection performance of the conventional Rao test [18] for the distributed target detection is also given, i.e.,

$$T_{Rao} = \frac{\sum_{h=1}^{H} \left| \boldsymbol{t}^H \hat{\boldsymbol{S}}^{-1} \boldsymbol{y}_h \right|^2}{\boldsymbol{t}^H \hat{\boldsymbol{S}}^{-1} \boldsymbol{t} \cdot \sum_{h=1}^{H} \boldsymbol{y}_h^H \hat{\boldsymbol{S}}^{-1} \boldsymbol{y}_h} \underset{H_0}{\overset{H_1}{\gtrless}} \eta_{Rao} \tag{2.38}$$

where $\hat{\boldsymbol{S}} = \sum_{h=H+1}^{H+K} \boldsymbol{y}_h \boldsymbol{y}_h^H$.

In Figure 2.3(a) and (b), the detection performance of the MRao detector and the conventional Rao detector is plotted for different K when $N = 10$ and $H = 3$. It can be seen that the MRao detector obtains higher PD than the conventional Rao test. The detection performance gains of the MRao detector with respect to the Rao detector are 5dB and 2dB for $K = 20$ and $K = 40$, respectively. Moreover, the detection performance improvement is significant when the training data are insufficient. Thus, the exploitation of the prior information of the noise can improve the detection performance effectively.

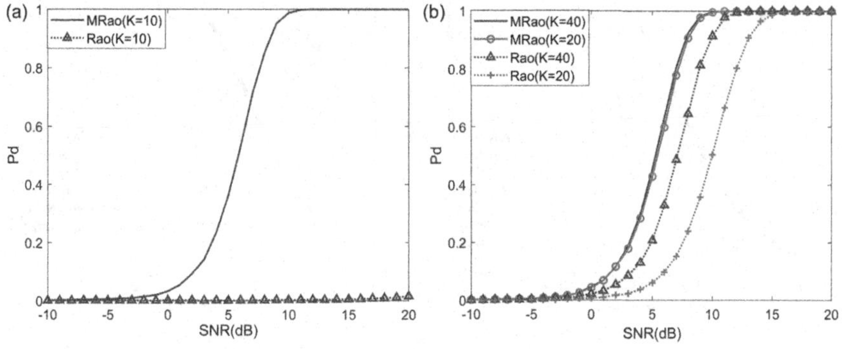

FIGURE 2.3 PD versus SNR for $N = 10$, $H = 3$. (a) $K = 10$; (b) $K = 20, 40$.

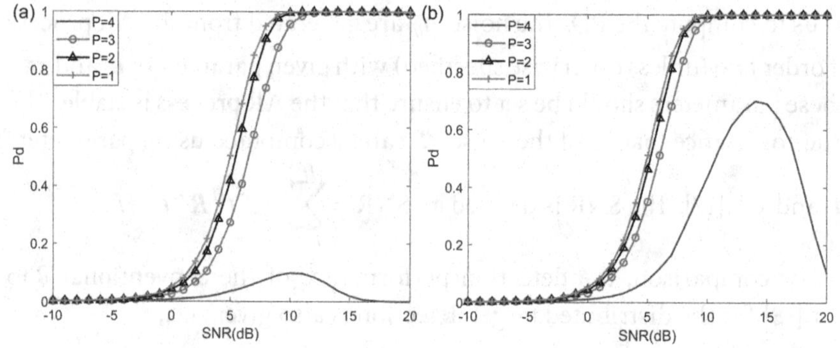

FIGURE 2.4 PD versus SNR for various P. (a) $K = 20$. (b) $K = 40$.

In Figure 2.4 (a) and (b), the PD of the MRao detector is plotted as a function of SNR for different P. From Figure 2.4, we can see that the detection performance of the MRao detector becomes worse as P increases. This is due to the fact that the estimated parameters may be incorrect when P samples have been discarded.

To analyze the influence of the pulse number N on the detection performance of the MRao detector, we plot the PD of the MRao versus SNR for different N in Figure 2.5. The asymptotic performance of the MRao is also given for comparison. It can be seen that the detection performance of the MRao increases with the increase of pulse number N. Meanwhile, even though the number of the training data is limited, the detection performance of the MRao detector approaches asymptotic performance when $N > 40$.

To see the influence of the H on the detection performance of the MRao detector, the PD of the MRao detector versus SNR for various H is

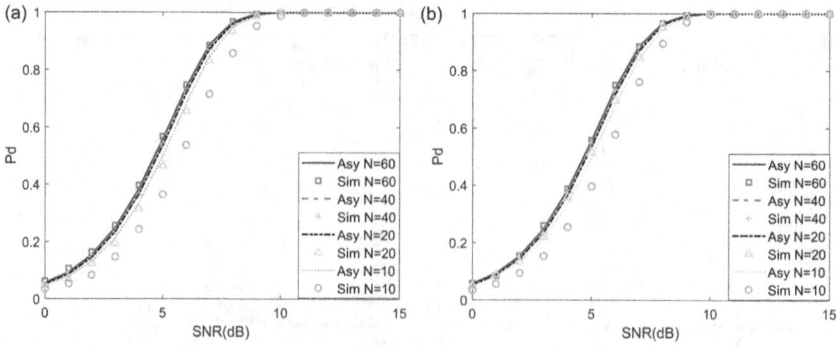

FIGURE 2.5 PD versus SNR for various N. (a) $K = 10$. (b) $K = 20$.

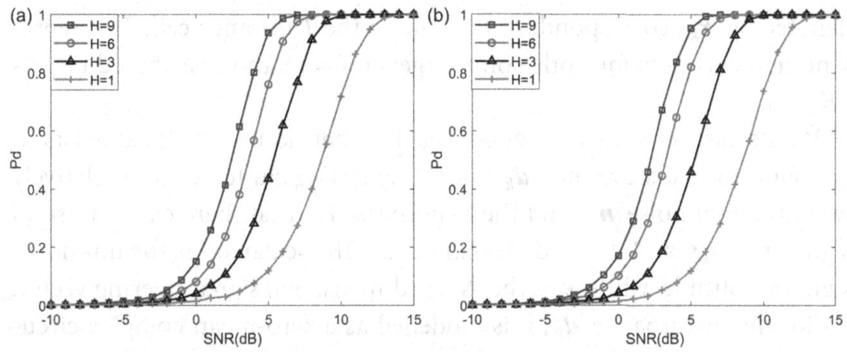

FIGURE 2.6 PD versus SNR for various H. (a) $K = 20$. (b) $K = 40$.

plotted in Figure 2.6. It can be seen that the increase of radar resolution or the range cells the target occupies can produce a detection performance gain.

2.3 ADAPTIVE DETECTOR BASED ON ORTHOGONAL PARTITION OF THE DATA

In this section, we design an adaptive detector based on orthogonal partition of the data in a sample-starved environment.

As pointed out in [19], the PHE model is well-suited for airborne radar systems with a moderately low number of primary and training data. Building upon the concept presented in [20], our research extends this approach to address both point-like and distributed target detection in PHE. The proposed detector utilizes a partitioning strategy on the primary and training data through orthogonal projection. This results in the separation of the clutter subspace and the clutter-free subspace. By projecting the primary data onto the clutter-free subspace, we estimate the power of the target. Similarly, the projection of the training data onto the clutter-free subspace allows us to estimate the residual noise within the target subspace. To achieve a CFAR property, we calculate the ratio between the estimator of the target's power and the estimator of the residual noise's power. This ratio is then divided by a generalized guard channel. The new detector provides higher PD compared to its natural competitors.

2.3.1 Problem Formulation

We consider a scenario where the target is present across K successive range cells. In the cell under test (CUT), the primary data are represented by an $N \times K$-dimensional matrix $\boldsymbol{X}$, where the kth column vector, denoted by $\boldsymbol{x}_k$, corresponds to the data in the kth range cell. This representation accounts for both point targets ($K=1$) and distributed targets ($K>1$).

We decide between the hypothesis H_0 that no useful signal exists in $\boldsymbol{x}_k$ but for the disturbance $\boldsymbol{d}_k$, including strong clutter $\boldsymbol{c}_k$ and relatively weak thermal noise $\boldsymbol{n}_k$, and the hypothesis H_1 that there exists a useful signal $\boldsymbol{s}_k = a_k\boldsymbol{s}$ and $K>1$ disturbance $\boldsymbol{d}_k$. The scalar a_k is the unknown signal amplitude, while $\boldsymbol{s}$ is the $N\times 1$-dimensional signal steering vector. As for the disturbance $\boldsymbol{d}_k$, it is modelled as a zero-mean complex circular Gaussian vector, with an unknown covariance matrix $\boldsymbol{R}_t$, written as $\boldsymbol{d}_k \sim \mathcal{CN}_N(\boldsymbol{0},\boldsymbol{R}_t)$. To estimate the unknown $\boldsymbol{R}_t$, we use L training data, denoted by an $N\times L$-dimensional matrix $\boldsymbol{X}_L = [\boldsymbol{x}_{s,1},\boldsymbol{x}_{s,2},\ldots,\boldsymbol{x}_{s,L}]$, where $\boldsymbol{x}_{s,l}$, only contains disturbance $\boldsymbol{d}_{s,l}$, with the covariance matrix $\boldsymbol{R}$. In the PHE, $\boldsymbol{R}_t$ shares the same structure with $\boldsymbol{R}$, but with an unknown power mismatch. Hence, we can express $\boldsymbol{R}_t$ as $\boldsymbol{R}_t = \sigma^2\boldsymbol{R}$, where σ^2 is an unknown power mismatch.[1]

To sum up, the detection problem to be solved can be formulated by the following binary hypothesis testing:

$$\begin{cases} \text{H}_0: \begin{cases} \boldsymbol{x}_k = \boldsymbol{d}_k,\ k=1,2,\ldots,K \\ \boldsymbol{x}_{s,l} = \boldsymbol{d}_{s,l},\ l=1,2,\ldots,L \end{cases} \\ \text{H}_1: \begin{cases} \boldsymbol{x}_k = a_k\boldsymbol{s} + \boldsymbol{d}_k,\ k=1,2,\ldots,K \\ \boldsymbol{x}_{s,l} = \boldsymbol{d}_{s,l},\ l=1,2,\ldots,L \end{cases} \end{cases} \tag{2.39}$$

In many situations, the disturbance $\boldsymbol{d}_k$ is composed of strong clutter and weak thermal noise. Then, the structure of the covariance matrix $\boldsymbol{R}$ can be factored into [21]

$$\boldsymbol{R} = \boldsymbol{R}_c + \sigma_n^2\boldsymbol{I}_N = \boldsymbol{Q}_c\Lambda_c\boldsymbol{Q}_c^H + \sigma_n^2\boldsymbol{I}_N \tag{2.40}$$

where $\boldsymbol{R}_c = \boldsymbol{Q}_c\Lambda_c\boldsymbol{Q}_c^H$ is the covariance matrix of the clutter, $\sigma_n^2\boldsymbol{I}_N$ is the covariance matrix of the thermal noise, and Λ_c is an $r\times r$-dimension diagonal matrix with eigenvalues of the clutter on its diagonal, which can be ordered in a non-decreasing manner, i.e., $\lambda_1 \geq \lambda_2 \geq \cdots \geq \lambda_r \gg \sigma_n^2$. σ_n^2 is the power of the thermal noise, which can be obtained by measuring receiver noise when the microwave radar does not transmit signals. For convenience, σ_n^2 is set to be one, i.e., $\sigma_n^2 = 1$. The value of r is defined as the clutter rank. $\boldsymbol{Q}_c$ is an $N \times r$-dimension semi-unitary matrix which consists of eigenvectors of the clutter and spans the clutter subspace, satisfying the equality $\boldsymbol{Q}_c^H \boldsymbol{Q}_c = \boldsymbol{I}_r$. Equation (2.40) can be expressed as

$$\boldsymbol{R} = \boldsymbol{Q}_c \Lambda_{cn} \boldsymbol{Q}_c^H + \boldsymbol{Q}_n \boldsymbol{Q}_n^H \tag{2.41}$$

where $\Lambda_{cn} = \Lambda_c + \boldsymbol{I}_r$, $\boldsymbol{Q}_n^H \boldsymbol{Q}_n = \boldsymbol{I}_{N-r}$ and $\boldsymbol{Q}_n^H \boldsymbol{Q}_c = \boldsymbol{0}_{(N-r)\times r}$. Hence we have the result $\boldsymbol{R}^{-1} = \boldsymbol{Q}_c \Lambda_{cn}^{-1} \boldsymbol{Q}_c^H + \boldsymbol{Q}_n \boldsymbol{Q}_n^H$. Since the clutter-to-noise ratio (CNR) can be quite high – up to 30 dB or more [20], $\boldsymbol{R}^{-1}$ is approximated by [21]

$$\boldsymbol{R}^{-1} \approx \boldsymbol{P}_n \tag{2.42}$$

where $\boldsymbol{P}_n = \boldsymbol{Q}_n \boldsymbol{Q}_n^H$ is the orthogonal projection matrix of the subspace orthogonal to the clutter subspace, with the rank $N-r$. In a similar manner, the orthogonal projection matrix of the clutter subspace can be formed as $\boldsymbol{P}_c = \boldsymbol{Q}_c \boldsymbol{Q}_c^H$, with the rank r. It can be easily shown that $\boldsymbol{P}_c^{\perp} = \boldsymbol{P}_n$.

2.3.2 Detector Design

The joint PDF of $\boldsymbol{X}$ and $\boldsymbol{X}_L$, under hypothesis H_1, is found to be

$$f_1(\boldsymbol{X}, \boldsymbol{X}_L) = \frac{\exp\left[-\mathrm{tr}(\boldsymbol{R}^{-1}\boldsymbol{S}_s) - \mathrm{tr}(\boldsymbol{R}^{-1}\boldsymbol{S}_p)/\sigma^2\right]}{\pi^{N(K+L)}\sigma^{2NK}\left|\boldsymbol{R}\right|^{K+L}} \tag{2.43}$$

where

$$\boldsymbol{S}_s = \boldsymbol{X}_L \boldsymbol{X}_L^H \tag{2.44}$$

$$\boldsymbol{S}_p = (\boldsymbol{X} - \boldsymbol{s}\boldsymbol{a}^H)(\boldsymbol{X} - \boldsymbol{s}\boldsymbol{a}^H)^H \tag{2.45}$$

and $\boldsymbol{a} = [a_1, a_2, \ldots, a_K]^T$. Note that the joint PDF of $\boldsymbol{X}$ and $\boldsymbol{X}_L$, under hypothesis H_0, also has a same form as Equation 2.45, but with $\boldsymbol{S}_p = \boldsymbol{X}\boldsymbol{X}^H$.

A well-known detector for the detection problem in Equation 2.39, is the generalized adaptive subspace detector (GASD) [19], whose detection statistic is given by

$$t_{\mathrm{GASD}} = \frac{\boldsymbol{s}^H \boldsymbol{S}_s^{-1} \boldsymbol{X}\boldsymbol{X}^H \boldsymbol{S}_s^{-1} \boldsymbol{s}}{\mathrm{tr}(\boldsymbol{X}^H \boldsymbol{S}_s^{-1} \boldsymbol{X}) \cdot \boldsymbol{s}^H \boldsymbol{S}_s^{-1} \boldsymbol{s}} \tag{2.46}$$

Equation (2.46) can be recast as

$$t_{\text{GASD}} = \frac{\operatorname{tr}\left(\tilde{X}^H P_{\tilde{s}} \tilde{X}\right)}{\operatorname{tr}\left(\tilde{X}^H \tilde{X}\right)} \tag{2.47}$$

where $P_{\tilde{s}} = \tilde{s}\tilde{s}^H / \tilde{s}^H \tilde{s}$ is the orthogonal projection matrix of $\tilde{s} = S_s^{-1/2} s$, $S_s^{-1/2} = (S_s^{1/2})^{-1}$, and $S_s^{1/2}$ is the square-root matrix of S_s.

When $K = 1$, X becomes a column vector, say x. Then the GASD degenerates into the famous adaptive coherence estimator (ACE) [22], described as

$$t_{\text{ACE}} = \frac{\tilde{x}^H P_{\tilde{s}} \tilde{x}}{\tilde{x}^H \tilde{x}} \tag{2.48}$$

or in another equivalent form

$$t_{\text{ACE}} = \frac{\left| s^H S_s^{-1} x \right|^2}{s^H S_s^{-1} s \cdot x^H S_s^{-1} x} \tag{2.49}$$

Adopting the method in [20], we project the primary and training data onto four independent subspaces, namely, the clutter subspace with the projection of the target being removed, the clutter subspace on which the target is projected, the clutter-free subspace with the projection of the target being removed, and the clutter-free subspace on which the target is projected. A large class of effective detectors is the so-called energy detector, which is essentially the ratio of the signal power to that of the residual noise in the target subspace [20].

Because the noise components in the clutter-free subspace are white and have the same power, all of the target-free data including the primary and training data in this clutter-free subspace can be used to estimate the noise power. According to the guideline mentioned above, the primary and training data can be partitioned into the eight parts [20]: $\mathcal{A}: P_s^{\perp} P_c X$, $\mathcal{B}: P_s^{\perp} P_c X_L$, $\mathcal{C}: P_s P_c X$, $\mathcal{D}: P_s P_c X_L$, $\mathcal{E}: P_s P_n X$, $\mathcal{F}: P_s P_n X_L$, $\mathcal{G}: P_s^{\perp} P_n X$, and $\mathcal{H}: P_s^{\perp} P_n X_L$. Note that $\mathcal{A}$, $\mathcal{B}$, $\mathcal{C}$, and $\mathcal{D}$ represent the components of the data related to the clutter subspace, $\mathcal{E}$, $\mathcal{F}$, $\mathcal{G}$, and $\mathcal{H}$ denote the components of the data related to the clutter-free subspace, and $\mathcal{C}$, $\mathcal{D}$, $\mathcal{E}$, and $\mathcal{F}$ stand for the components of the data related to the target subspace. A few remarks are in order. First, although the dimension of the target subspace is one, the target vector can be partitioned into several parts. Second, in general, the target vector is not entirely in the clutter subspace or entirely in the clutter-free subspace. Otherwise stated, the components of the projection of the target vector onto the clutter and clutter-free subspaces are both not empty in general. In fact, if the target vector entirely lies in the clutter subspace the target can never be detected. In contrast, if the target vector entirely lies in the clutter-free subspace there is no need for clutter suppression.

In [20], $\mathcal{F}$, $\mathcal{G}$, and $\mathcal{H}$ are used to form the MLE of the power of the residual noise based on the assumption of homogeneity. This estimator, however, is not the MLE of the power of the residual noise in PHE. Furthermore, it is biased in this scenario. In fact, the MLE of the power of the residual noise in PHE is

$$\hat{\sigma}_r^2 = \mathrm{tr}\left(\boldsymbol{X}_L^H \boldsymbol{P}_n \boldsymbol{X}_L\right) / N_s' \tag{2.50}$$

where $N_s' = (N - r)L$. Using the guideline mentioned in the beginning of this section, we can form a detector as

$$t_1 = \frac{\hat{E}_t^2}{\hat{\sigma}_r^2} = \frac{\mathrm{tr}\left(\boldsymbol{X}^H \boldsymbol{P}_n \boldsymbol{s}\boldsymbol{s}^H \boldsymbol{P}_n \boldsymbol{X}\right) / \boldsymbol{s}^H \boldsymbol{P}_n \boldsymbol{s}}{\mathrm{tr}\left(\boldsymbol{X}_L^H \boldsymbol{P}_n \boldsymbol{X}_L\right) / N_s'} \tag{2.51}$$

where $\hat{E}_t^2$, implicitly defined in (2.51), is the target power in the target subspace with the clutter being suppressed. However, the detector in Equation 2.51 is not CFAR in PHE. Some modifications are needed to maintain CFAR property. It is helpful to review the normalized matched filter (NMF) for the point target with the following statistic

$$t_{\mathrm{NMF}} = \frac{|\boldsymbol{s}^H \boldsymbol{R}^{-1} \boldsymbol{x}|^2}{\boldsymbol{s}^H \boldsymbol{R}^{-1} \boldsymbol{s} \cdot \boldsymbol{x}^H \boldsymbol{R}^{-1} \boldsymbol{x}} = \frac{t_{\mathrm{MF}}}{\boldsymbol{x}^H \boldsymbol{R}^{-1} \boldsymbol{x}} \tag{2.52}$$

where $t_{\mathrm{MF}} = |\boldsymbol{s}^H \boldsymbol{R}^{-1} \boldsymbol{x}|^2 / \boldsymbol{s}^H \boldsymbol{R}^{-1} \boldsymbol{s}$ is the optimum detector for the point target in homogeneous environments, with the name MF [23]. On the other hand, the NMF is an optimum CFAR detector for the point target in PHE. It is seen that the NMF is the ratio of the MF to the quantity $\boldsymbol{x}^H \boldsymbol{R}^{-1} \boldsymbol{x}$, which is in the form of the generalized inner product, and it is called a generalized guard channel in [24]. Thus, one can divide Equation 2.51 by the quantity $\mathrm{tr}\left(\boldsymbol{X}^H \boldsymbol{R}^{-1} \boldsymbol{X}\right)$, i.e., the entire power of the primary data after whitening, to obtain a CFAR detector in PHE for the distributed target. Since $\boldsymbol{R}^{-1} \approx \boldsymbol{P}_n$ for high CNR scenario, we use the quantity $\mathrm{tr}\left(\boldsymbol{X}^H \boldsymbol{P}_n \boldsymbol{X}\right)$ instead of $\mathrm{tr}\left(\boldsymbol{X}^H \boldsymbol{R}^{-1} \boldsymbol{X}\right)$. Therefore, the proposed CFAR detector in PHE is given by

$$t_{\text{OP-RR-GNMF}} = \frac{\mathrm{tr}\left(\boldsymbol{X}^H \boldsymbol{P}_{\boldsymbol{P}_n \boldsymbol{s}} \boldsymbol{X}\right)}{\mathrm{tr}\left(\boldsymbol{X}^H \boldsymbol{P}_n \boldsymbol{X}\right) \cdot \mathrm{tr}\left(\boldsymbol{X}_L^H \boldsymbol{P}_n \boldsymbol{X}_L\right)} \tag{2.53}$$

where

$$\boldsymbol{P}_{\boldsymbol{P}_n \boldsymbol{s}} = \frac{\boldsymbol{P}_n \boldsymbol{s} \boldsymbol{s}^H \boldsymbol{P}_n}{\boldsymbol{s}^H \boldsymbol{P}_n \boldsymbol{s}} \tag{2.54}$$

is the orthogonal projection matrix of $\boldsymbol{P}_n \boldsymbol{s}$. For convenience, we denote the detector in (2.53) as the orthogonal-partition reduced-rank generalized NMF (OP-RR-GNMF).

Note that when $K = 1$, (2.53) reduces to

$$t_{\text{OP-RR-NMF}} = \frac{\boldsymbol{x}^H \boldsymbol{P}_{\boldsymbol{P}_n \boldsymbol{s}} \boldsymbol{x}}{\mathrm{tr}\left(\boldsymbol{X}_L^H \boldsymbol{P}_n \boldsymbol{X}_L\right) \cdot \boldsymbol{x}^H \boldsymbol{P}_n \boldsymbol{x}} \tag{2.55}$$

which is the counterpart of Equation (2.53) for the point target detection, and is referred to as the orthogonal-partition reduced-rank NMF (OP-RR-NMF).

In conclusion, the procedure of the design of the orthogonal-partition reduced-rank detector can be summarized in the following five steps: 1) compute the MLE of the clutter power in the signal subspace, 2) calculate the signal power in the signal subspace when the clutter is suppressed, 3) calculate the residual noise power in the signal subspace when the clutter is suppressed, 4) calculate the ratio of the term of the step 2) to that of the step 3), 5) divide the ratio of 4) by the generalized guard channel and compare it with the detection threshold.

In the above derivations, we assume that the covariance matrix is known. In the case of an unknown covariance matrix, we need training data to estimate the unknown covariance matrix. It is well known that the MLE of $\boldsymbol{R}$ is the sample covariance matrix, formed as

$$\hat{\boldsymbol{R}} = \frac{1}{L} \sum_{l=1}^{L} \boldsymbol{x}_{s,l} \boldsymbol{x}_{s,l}^H \tag{2.56}$$

In a manner analogous to Equation 2.41, $\hat{R}$ can be factored into $\hat{R}=\hat{Q}_c\hat{\Lambda}_{cn}\hat{Q}_c^H+\hat{Q}_n\hat{Q}_n^H$. Hence the estimator of P_n can be obtained as $\hat{P}_n=\hat{Q}_n\hat{Q}_n^H$. Therefore, when P_n is unknown we can modify Equation 2.53 by replacing P_n with $\hat{P}_n$, namely,

$$t_{\text{OP-RR-GANMF}}=\frac{\mathrm{tr}\left(X^H P_{\hat{P}_n s}X\right)}{\mathrm{tr}\left(X^H\hat{P}_nX\right)\cdot\mathrm{tr}\left(X_L^H\hat{P}_nX_L\right)} \tag{2.57}$$

which is referred to as the orthogonal-partition reduced-rank generalized ANMF (OP-RR-GANMF).

When $K=1$, Equation 2.57 becomes

$$t_{\text{OP-RR-ANMF}}=\frac{x^H P_{\hat{P}_n s}x}{x^H\hat{P}_nx\cdot\mathrm{tr}\left(X_L^H\hat{P}_nX_L\right)} \tag{2.58}$$

which is denoted as the orthogonal-partition reduced-rank ANMF (OP-RR-ANMF).

2.3.3 Performance Assessment

It is difficult to obtain the exact statistical property of the OP-RR-GANMF. Hence, we only analyze the statistical properties of the OP-RR-GNMF. For convenience, we introduce the following two notations

$$t_{nuw}=\frac{\mathrm{tr}\left(X^H P_{P_n s}X\right)}{\mathrm{tr}\left(X^H P_nX\right)} \tag{2.59}$$

and

$$t_{den}=\mathrm{tr}\left(X_L^H P_nX_L\right) \tag{2.60}$$

Then Equation 2.53 can be written as

$$t_{\text{OP-RR-GNMF}}=\frac{t_{num}}{t_{den}} \tag{2.61}$$

Equation 2.59 can be expressed as

$$t_{nuw}=\frac{\mathrm{tr}\left(\bar{X}^H R^{1/2}P_{P_n s}R^{1/2}\bar{X}\right)}{\mathrm{tr}\left(\bar{X}^H R^{1/2}P_nR^{1/2}\bar{X}\right)} \tag{2.62}$$

where $\bar{\boldsymbol{X}} = \boldsymbol{R}^{-1/2}\boldsymbol{X}$, $\boldsymbol{R}^{-1/2} = (\boldsymbol{R}^{1/2})^{-1}$, and $\boldsymbol{R}^{1/2}$ is square-root matrix of $\boldsymbol{R}$. The square-root matrix is not unique, and a common one is given by

$$\boldsymbol{R}^{1/2} = (\boldsymbol{Q}_c\boldsymbol{\Lambda}_c^{1/2}\boldsymbol{Q}_c^H + \boldsymbol{I}_N) \tag{2.63}$$

One can readily verify the following two equations hold [25]:

$$\boldsymbol{P}_n\boldsymbol{R}\boldsymbol{P}_n = \boldsymbol{P}_n \tag{2.64}$$

$$\boldsymbol{R}^{1/2}\boldsymbol{P}_{\boldsymbol{P}_n s}\boldsymbol{R}^{1/2} = \boldsymbol{P}_{\boldsymbol{P}_n s} \tag{2.65}$$

According to Equations 2.64 and 2.65, Equation 2.59 can be recast as

$$t_{num} = \frac{\mathrm{tr}\left(\bar{\boldsymbol{X}}^H\boldsymbol{P}_{\boldsymbol{P}_n s}\bar{\boldsymbol{X}}\right)}{\mathrm{tr}\left(\bar{\boldsymbol{X}}^H\boldsymbol{P}_n\bar{\boldsymbol{X}}\right)} \tag{2.66}$$

Taking a similar manner, we can rewrite Equation 2.60 as

$$t_{den} = \mathrm{tr}\left(\bar{\boldsymbol{X}}_L^H\boldsymbol{P}_n\bar{\boldsymbol{X}}_L\right) \tag{2.67}$$

where $\bar{\boldsymbol{X}}_L = \boldsymbol{R}^{-1/2}\boldsymbol{X}_L$.

Moreover, Equations 2.66 and 2.67 can be further expressed as

$$t_{num} = \frac{\sum_{k=1}^{K}\bar{\boldsymbol{x}}_k^H\boldsymbol{P}_{\boldsymbol{P}_n s}\bar{\boldsymbol{x}}_k \Big/ \sigma^2}{\sum_{k=1}^{K}\bar{\boldsymbol{x}}_k^H\boldsymbol{P}_n\bar{\boldsymbol{x}}_k \Big/ \sigma^2} \tag{2.68}$$

and

$$t_{den} = \sum_{l=1}^{L}\bar{\boldsymbol{x}}_{s,l}^H\boldsymbol{P}_n\bar{\boldsymbol{x}}_{s,l} \tag{2.69}$$

respectively.

In (2.68) and (2.69), $\bar{\boldsymbol{x}}_k = \boldsymbol{R}^{-1/2}\boldsymbol{x}_k$ and $\bar{\boldsymbol{x}}_{s,l} = \boldsymbol{R}^{-1/2}\boldsymbol{x}_{s,l}$, where $\boldsymbol{x}_k$ is the k th column of the primary data $\boldsymbol{X}$ and $\boldsymbol{x}_{s,l}$ is the l th column of the training data $\boldsymbol{X}_L$.

It is easy to show that $\bar{x}_k$ under hypothesis H_1 is distributed as $\bar{x}_k \sim \mathcal{CN}_N(\bar{\mu}, \sigma^2 I_N)$, with $\bar{\mu} = a_k R^{-1/2} s$. It follows that as $\bar{x}_k/\sigma \sim \mathcal{CN}_N(\bar{\mu}/\sigma, I_N)$. Moreover, $\bar{x}_{s,l}$, both under hypotheses H_1 and H_0, is distributed as $\bar{x}_{s,l} \sim \mathcal{CN}_N(\mathbf{0}, I_N)$.

Equation (2.68) can be represented by

$$t_{num} = \frac{\sum_{k=1}^{K} \bar{x}_k^H P_{P_n s} \bar{x}_k \Big/ \sigma^2}{\sum_{k=1}^{K} \bar{x}_k^H P_{P_n s} \bar{x}_k \Big/ \sigma^2 + \sum_{k=1}^{K} \bar{x}_k^H P_n^{\bar{s}} \bar{x}_k \Big/ \sigma^2} \tag{2.70}$$

where

$$P_n^{\bar{s}} = P_n - P_{P_n s} \tag{2.71}$$

It can be verified that $P_n^{\bar{s}}$ is an $N \times N$-dimensional orthogonal projection matrix of rank $(N - r - 1)$. Thereby, $\bar{x}_k^H P_n^{\bar{s}} \bar{x}_k$ is distributed as $\bar{x}_k^H P_n^{\bar{s}} \bar{x}_k \sim \mathcal{C}\chi^2_{N-r-1}(\rho_0)$ [26]. It is observed that $\rho_0 = \bar{\mu}^H P_n^{\bar{s}} \bar{\mu} = 0$. Hence, both under hypotheses H_1 and H_0, $\bar{x}_k^H P_n^{\bar{s}} \bar{x}_k$ is distributed as

$$\bar{x}_k^H P_n^{\bar{s}} \bar{x}_k \sim \mathcal{C}\chi^2_{N-r-1} \tag{2.72}$$

Taking a manner analogous to that of Equation 2.72, one can verify that under hypothesis H_1

$$\bar{x}_k^H P_{P_n s} \bar{x}_k / \sigma^2 \sim \mathcal{C}\chi^2_1(\rho_{r,k}) \tag{2.73}$$

where

$$\rho_{r,k} = |a_k|^2 \cdot s^H P_n s \tag{2.74}$$

In Equation 2.74, we have used the fact that

$$s^H R^{-1/2} P_{P_n s} R^{-1/2} s = s^H P_n s \tag{2.75}$$

Note that (2.74) can be further expressed as $\rho_{r,k} = |a_k|^2 \|P_n s\|^2 = |a_k|^2 \cdot \|P_c^{\perp} s\|^2$, where $\|P_c^{\perp} s\|^2$ can be taken as the projection of the signal energy in the subspace which is orthogonal to the clutter subspace. It is well known that

$$\rho = |a|^2 \boldsymbol{s}^H \boldsymbol{R}_t^{-1} \boldsymbol{s} \tag{2.76}$$

is the optimum output signal-to-clutter-plus-noise ratio (SCNR) of the MF for the case of $K = 1$. The value of ρ_r is smaller than that of ρ. However, under the scenario where the CNR is high enough, ρ_r approaches ρ, i.e., $\rho_r \approx \rho$ [25].

Since $\bar{\boldsymbol{x}}_k$ is statistically independent of each other for $k = 1,2,\ldots,K$, and given the considerations outlined in (2.73), the numerator of (2.70) is ruled by

$$\sum_{k=1}^{K} \bar{\boldsymbol{x}}_k^H \boldsymbol{P}_{\boldsymbol{P}_n \boldsymbol{s}} \bar{\boldsymbol{x}}_k \Big/ \sigma^2 \sim \mathcal{C}\chi_K^2(\rho_r) \tag{2.77}$$

where $\rho_r = \sum_{k=1}^{K} \rho_{r,k}$.

Similarly, according to (2.72), the quantity $\sum_{k=1}^{K} \bar{\boldsymbol{x}}_k^H \boldsymbol{P}_n^{\bar{\boldsymbol{s}}} \bar{\boldsymbol{x}}_k \Big/ \sigma^2$ in the denominator of (2.70) is subject to

$$\sum_{k=1}^{K} \bar{\boldsymbol{x}}_k^H \boldsymbol{P}_n^{\bar{\boldsymbol{s}}} \bar{\boldsymbol{x}}_k \Big/ \sigma^2 \sim \mathcal{C}\chi_{K(N-r-1)}^2 \tag{2.78}$$

We observe that the orthogonal projection $\boldsymbol{P}_n^{\bar{\boldsymbol{s}}}$ in (2.71) is orthogonal to $\boldsymbol{P}_{\boldsymbol{P}_n \boldsymbol{s}}$, i.e., $\boldsymbol{P}_n^{\bar{\boldsymbol{s}}} \boldsymbol{P}_{\boldsymbol{P}_n \boldsymbol{s}} = \boldsymbol{P}_{\boldsymbol{P}_n \boldsymbol{s}} \boldsymbol{P}_n^{\bar{\boldsymbol{s}}} = \boldsymbol{0}$. Hence, the term $\bar{\boldsymbol{x}}_k^H \boldsymbol{P}_{\boldsymbol{P}_n \boldsymbol{s}} \bar{\boldsymbol{x}}_k$ is statistically independent of $\bar{\boldsymbol{x}}_k^H \boldsymbol{P}_n^{\bar{\boldsymbol{s}}} \bar{\boldsymbol{x}}_k$. It follows that t_{num} in (2.70), under hypothesis H_1, is statistically equivalent to

$$t_{num} \overset{d}{=} 1 - \beta_{num} \tag{2.79}$$

where β_{num} is distributed as $\beta_{num} \sim \mathcal{C}\beta_{K,K(N-r-1)}(\rho_r)$. A noncentral complex Beta random variable β, with M_1 and M_2 DOFs and a noncentrality parameter ρ is defined as

$$\beta = c_2 / (c_1 + c_2) \tag{2.80}$$

where $c_1 \sim \mathcal{C}\chi_{M_1}^2(\rho_r)$ and $c_2 \sim \mathcal{C}\chi_{M_2}^2$.

Now we proceed to derive the statistical distribution of the t_{den}. From (2.69), it can be achieved that under both hypotheses H_1 and H_0, the statistical distribution of t_{den} is

$$t_{den} \sim C\chi^2_{L(N-r)} \tag{2.81}$$

Therefore, $t_{\text{OP-RR-GNMF}}$ is statistically equivalent to the ratio of (2.79) to (2.81), described as

$$t_{\text{OP-RR-GNMF}} \overset{d}{=} \frac{1 - C\beta_{K,K(N-r-1)}(\rho_r)}{C\chi^2_{L(N-r)}} \tag{2.82}$$

It is worth pointing out that when $K = 1$, i.e., the case of point target, (2.82) reduces to

$$t_{\text{OP-RR-GNMF}} \overset{d}{=} \frac{1 - C\beta_{1,N-r-1}(\rho_r)}{C\chi^2_{L(N-r)}} \tag{2.83}$$

It is difficult to derive the PDF of (2.82). However, (2.82) is useful at least in the following two facts: 1) the detector OP-RR-GNMF is strictly CFAR in PHE; 2) (2.82) is the stochastic representation of the OP-RR-GNMF, and hence the detection performance can be obtained by using (2.82) rather than using (2.53), which largely reduces the computational complexity because of no need of the matrix operation, especially the time-consuming matrix inversion.

2.3.4 Numerical Examples

In this section, we evaluate the detection performance of the orthogonal-partition detector. We consider a scenario of airborne radar. The radar array operates in side-looking mode and the clutter slope is set to be unity. The radar array has $N_a = 4$ antenna elements, and each antenna element receives $N_p = 4$ pulses. Hence, the rank of clutter is $r = N_a + N_p - 1 = 7$. The signal steering vector has the form $\boldsymbol{s} = \boldsymbol{s}_s \otimes \boldsymbol{s}_t$, where $\boldsymbol{s}_s$ and $\boldsymbol{s}_t$ are the space and time steering vectors, respectively, having the forms

$$\boldsymbol{s}_s = [1, e^{-j2\pi\theta}, \cdots, e^{-j2\pi(N_a-1)\theta}]^T \tag{2.84}$$

and

$$\boldsymbol{s}_t = [1, e^{-j2\pi f_d}, \cdots, e^{-j2\pi(N_p-1)f_d}]^T \tag{2.85}$$

with θ and f_d being the normalized spatial frequency and normalized Doppler frequency, respectively. We set $\theta = 0.1$ and $f_d = 0.25$. The CNR is defined as $\mathrm{CNR} = \sum_{i=1}^{r} \lambda_i / \sigma_n^2$, which is chosen as 60 dB. To alleviate the computational complexity, the PFA is chosen to be $\mathrm{PFA} = 10^{-3}$ through this subsection. The detection thresholds are obtained through 100/PFA Monte Carlo simulations, and the corresponding PDs are obtained using 10^4 data realizations.

The SCNR is defined as

$$\rho = \boldsymbol{a}^H \boldsymbol{a} \cdot \boldsymbol{s}^H \boldsymbol{R}_t^{-1} \boldsymbol{s} \tag{2.86}$$

Note that when $K = 1$, Equation 2.86 reduces to Equation 2.76.

For comparison purposes, we use the following detector

$$t_{\mathrm{GNMF}} = \frac{\boldsymbol{s}^H \boldsymbol{R}^{-1} \boldsymbol{X} \boldsymbol{X}^H \boldsymbol{R}^{-1} \boldsymbol{s}}{\mathrm{tr}\left(\boldsymbol{X}^H \boldsymbol{R}^{-1} \boldsymbol{X}\right) \cdot \boldsymbol{s}^H \boldsymbol{R}^{-1} \boldsymbol{s}} \tag{2.87}$$

as the benchmark detector. The detector in (2.87) is the GLRT for the problem in (2.39) when the covariance matrix $\boldsymbol{R}$ is known in advance. In fact, it is the distributed-target version of (2.52). For convenience, it is denoted as the generalized NMF (GNMF). The brief derivation of (2.87) is given in Appendix 2.B.

Figure 2.7 displays the detection performance of the OP-RR-GNMF, GASD, and GNMF. The results highlight that the PD of the OP-RR-GNMF is lower than that of the GNMF, but much higher than that of the GASD. Precisely, the performance improvement in terms of SCNR is 4 dB at PD=0.8. In fact, the performance improvement is much larger when the amount of training data is small, as shown below.

Figure 2.8 shows the PDs of the three detectors when the number of training data is slightly larger than the system dimension. The results indicate that both the OR-RR-GANMF and GASD suffer from certain performance losses, compared with the GNMF. However, the PD of the OR-RR-GANMF is much higher than that of the GASD.

Figure 2.9 depicts the PDs of the OP-RR-GNMF and GNMF when the number of training data is lower than the system dimension. In this sample-starved environment, the GASD is invalid. Hence, the PD of the GASD is not given in (2.9). It is seen that the PD of the OP-RR-GNMF

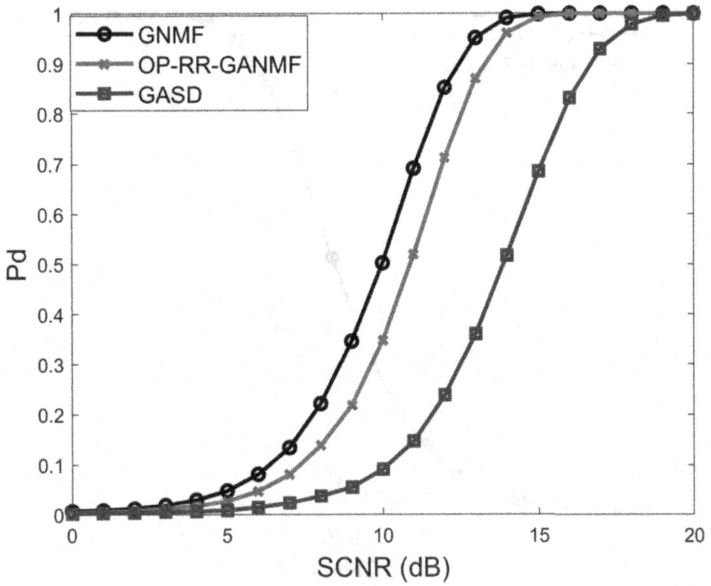

FIGURE 2.7 PDs of the OP-RR-GNMF, GASD, and GNMF versus SCNR. $L = 32$ and $K = 3$.

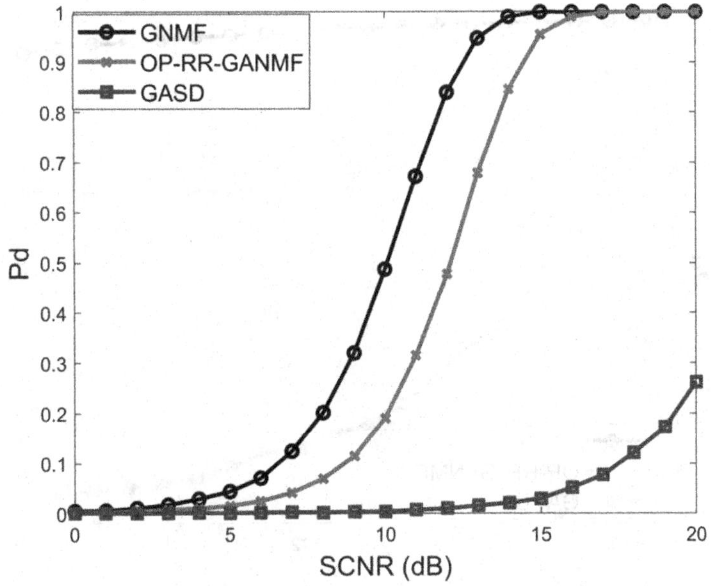

FIGURE 2.8 PDs of the OP-RR-GNMF, GASD, and GNMF versus SCNR. $L = 19$ and $K = 3$.

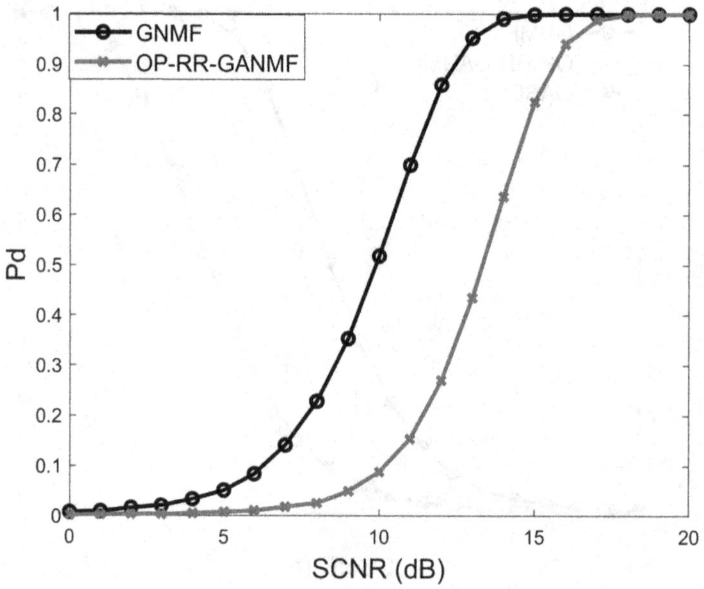

FIGURE 2.9 PDs of the OP-RR-GNMF, GASD, and GNMF versus SCNR. $L = 15$ and $K = 3$.

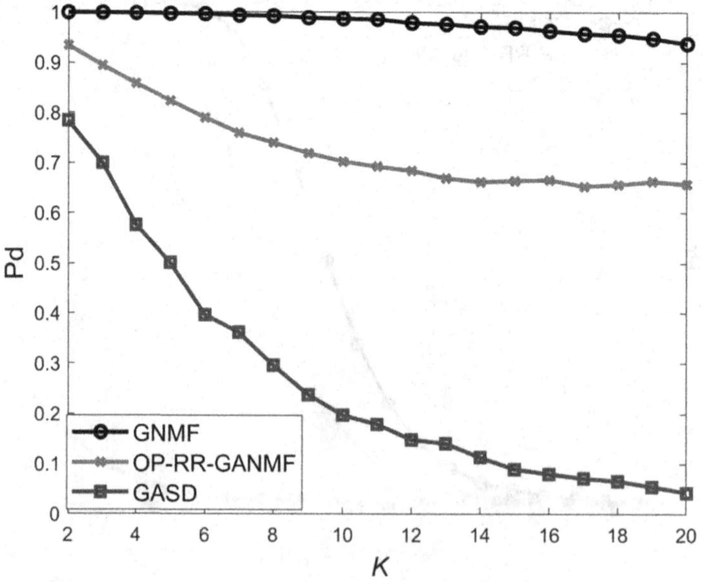

FIGURE 2.10 PDs of the OP-RR-GANMF, GASD, and GNMF versus K. $L = 32$ and $\text{SNR} = 15$.

decreases. Fortunately, the PD of the OR-RR-GNMF approaches unity when the SCNR is greater than 17 dB.

Figure 2.10 displays the PDs of the detectors under different K. It is seen that with the increasing of the value of K, the PDs of all detectors decrease. This is due to the fact that as K increases, the target energy becomes more dispersed, which is detrimental to detection performance.

2.4 ADAPTIVE DISTRIBUTED TARGET DETECTION IN THE PARTIALLY HOMOGENEOUS NOISE WITH PERSYMMETRIC STRUCTURE

In Section 2.1 and Section 2.2, adaptive detectors designed by modelling the disturbance as an AR process are given in the sample-starved environment. In this section, the approach that exploits the persymmetric structure [27–33] of the noise covariance matrix to alleviate the requirement of sufficient IID training data is introduced. In [30], a persymmetric one-step GLRT (P1SGLRT) and a persymmetric two-step GLRT (P2SGLRT) are proposed to detect distributed targets. However, the design of adaptive persymmetric detectors based on the Rao and Wald tests for distributed targets has not received much attention. Motivated by this, we derive the Rao and Wald tests for the problem of detecting distributed targets in noise with a persymmetric structure.

2.4.1 Problem Formulation

We assume that the primary data $y_t \in \mathbb{C}^{N\times 1}, t = 1,\ldots,L$ are received from N channels [34]. As shown in Figure 2.11, the target occupies L range cells. We denote the training data, free of target signal, by $y_t \in \mathbb{C}^{N\times 1}$, $t = L+1,\ldots,L+K$. Here, t denotes the index of the range cell. The detection problem is to decide whether $y_t \in \mathbb{C}^{N\times 1}, t = 1,\ldots,L$ contains target signals or not. We formulate the detection problem at hand as the following binary hypothesis testing

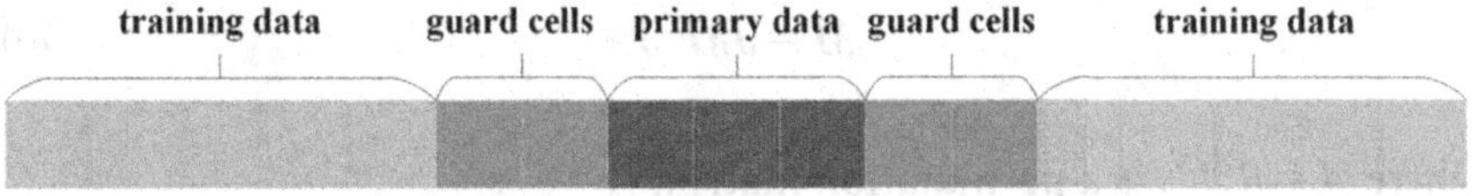

FIGURE 2.11 The primary data and training data of radar system.

$$\begin{cases} H_0: \boldsymbol{y}_t = \boldsymbol{n}_t, & t = 1,\ldots,L+K, \\ H_1: \begin{cases} \boldsymbol{y}_t = \alpha_t \boldsymbol{v} + \boldsymbol{n}_t, & t = 1,\ldots,L, \\ \boldsymbol{y}_t = \boldsymbol{n}_t, & t = L+1,\ldots L+K, \end{cases} \end{cases} \tag{2.88}$$

where $\boldsymbol{v} \in \mathbb{C}^{N\times 1}$ is the nominal steering vector, α_ts, $t = 1,\ldots,L$ are unknown deterministic parameters accounting for the target reflectivity and channel effects, and $\boldsymbol{n}_t \in \mathbb{C}^{N\times 1}, t = 1,\ldots,L+K$ are zero-mean complex Gaussian vectors.

In Section 2.1 and Section 2.2, it is assumed that the noise is homogeneous, namely, the training data are independent and identically distributed with the primary data. In practice, the assumption usually does not hold. The partially homogeneous noise is a widely used noise model, which has been verified by the real data collected from the Radar Surveillance Technology Experimental Radar system [35] and the over-the-horizon radar system [36]. In the partially homogeneous noise model, the training data share the same noise covariance matrix structure with the primary data up to an unknown scaling factor. That is to say,

$$\begin{cases} E\left[\boldsymbol{n}_t \boldsymbol{n}_t^H\right] = \boldsymbol{M}, & t = 1,\ldots,L, \\ E\left[\boldsymbol{n}_t \boldsymbol{n}_t^H\right] = \gamma \boldsymbol{M}, & t = L+1,\ldots,L+K, \end{cases} \tag{2.89}$$

where $\gamma > 0$ denotes the unknown power mismatch between the data under test and training data.

The persymmetric structure is exploited to design the adaptive detectors. The persymmetric structure exists when symmetrically spaced linear arrays and/or pulse trains are used. In addition, the persymmetry property can also be found in other situations, e.g., standard rectangular arrays, uniform cylindrical arrays (with an even number of elements), and some standard hexagonal arrays [27]. When the covariance matrix $\boldsymbol{M}$ is persymmetric, the elements of the covariance matrix are symmetric with respect to the main diagonal and Hermitian with respect to the auxiliary diagonal:

$$\boldsymbol{M} = \boldsymbol{J}\boldsymbol{M}^*\boldsymbol{J} \tag{2.90}$$

where $\boldsymbol{J} \in R^{N\times N}$ is a permutation matrix:

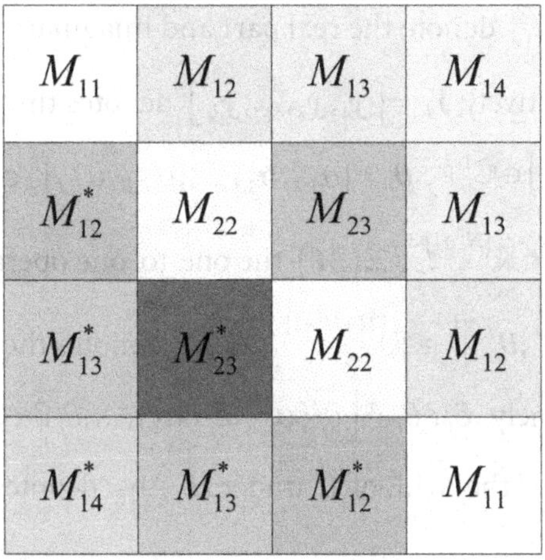

FIGURE 2.12 An illustration of a four-element persymmetric covariance matrix.

$$\boldsymbol{J} = \begin{bmatrix} 0 & 0 & \cdots & 0 & 1 \\ 0 & 0 & \cdots & 1 & 0 \\ \vdots & \vdots & \vdots & \vdots & \vdots \\ 1 & 0 & \cdots & 0 & 0 \end{bmatrix}$$

An illustration of a four-element persymmetric covariance matrix is shown in Figure 2.12. Meanwhile, it is assumed that the steering vector $\boldsymbol{v}$ satisfies $\boldsymbol{v} = \boldsymbol{J}\boldsymbol{v}^*$.

In the following subsections, the real parameter Rao test and real parameter Wald test, which divide the complex parameter into the real and imaginary parts, are resorted to solve the binary hypothesis testing problem in Equation 2.88.

2.4.2 The Rao Test

The real parameter Rao test is [13]

$$\left.\frac{\partial \ln f\left(\boldsymbol{y}_{1:L+K}|\boldsymbol{\theta},\boldsymbol{M}\right)}{\partial \boldsymbol{\theta}_r}\right|^T_{\boldsymbol{\theta}=\hat{\boldsymbol{\theta}}_0} \left[\boldsymbol{J}^{-1}\left(\hat{\boldsymbol{\theta}}_0\right)\right]_{\boldsymbol{\theta}_r,\boldsymbol{\theta}_r} \times \left.\frac{\partial \ln f\left(\boldsymbol{y}_{1:L+K}|\boldsymbol{\theta},\boldsymbol{M}\right)}{\partial \boldsymbol{\theta}_r}\right|_{\boldsymbol{\theta}=\hat{\boldsymbol{\theta}}_0} \underset{H_0}{\overset{H_1}{\gtrless}} \eta_R \quad (2.91)$$

where $\alpha_{t,R}$ and $\alpha_{t,I}$ denote the real part and imaginary part of α_t, $t=1,\ldots,L$, respectively, $\boldsymbol{Y}_L=\left[\boldsymbol{y}_1,\boldsymbol{y}_2,\ldots,\boldsymbol{y}_L\right]$ denotes the primary data, $\boldsymbol{\alpha}=\left[\alpha_1,\alpha_2,\ldots,\alpha_L\right]\in\mathbb{C}^{1\times L}$, $\boldsymbol{\theta}_r=[\alpha_{1,R},\alpha_{1,I},\ldots,\alpha_{L,R},\alpha_{L,I}]^T\in\mathbb{R}^{(2L)\times 1}$, $\boldsymbol{\theta}_s=[\gamma,\Xi(\boldsymbol{M})^T]^T\in\mathbb{R}^{(N^2+1)\times 1}$, $\Xi(\boldsymbol{M})$ the one-to-one operator mapping $\boldsymbol{M}$ to $\boldsymbol{\theta}_s$, $\boldsymbol{\theta}=\left[\boldsymbol{\theta}_r^T,\boldsymbol{\theta}_s^T\right]^T\in\mathbb{R}^{(2L+N^2+1)\times 1}$, $\partial/\partial\boldsymbol{\theta}_r$ denotes the gradient with respect to $\boldsymbol{\theta}_r$ (namely, $\partial/\partial\boldsymbol{\theta}_r=\left[\partial/\partial\alpha_{1,R},\partial/\partial\alpha_{1,I},\ldots,\partial/\partial\alpha_{L,R},\partial/\partial\alpha_{L,I}\right]^T$), $\hat{\boldsymbol{\theta}}_0=\left[\hat{\boldsymbol{\theta}}_{r,0}^T,\hat{\boldsymbol{\theta}}_{s,0}^T\right]^T$ is the MLE of $\boldsymbol{\theta}$ under H_0, η_R denotes the detection threshold, $\boldsymbol{J}(\boldsymbol{\theta})=\boldsymbol{J}(\boldsymbol{\theta}_r,\boldsymbol{\theta}_s)$ denotes the FIM and can be partitioned as [13] $\boldsymbol{J}(\boldsymbol{\theta})=\begin{bmatrix}\boldsymbol{J}_{\theta_r,\theta_r}(\boldsymbol{\theta}) & \boldsymbol{J}_{\theta_r,\theta_s}(\boldsymbol{\theta})\\ \boldsymbol{J}_{\theta_s,\theta_r}(\boldsymbol{\theta}) & \boldsymbol{J}_{\theta_s,\theta_s}(\boldsymbol{\theta})\end{bmatrix}$, where

$$\left[\boldsymbol{J}_{\theta_r,\theta_r}(\boldsymbol{\theta})\right]=-E\left[\frac{\partial^2\ln f\left(\boldsymbol{y}_{1:L+K}\,|\,\boldsymbol{\theta}\right)}{\partial\boldsymbol{\theta}_r\partial\boldsymbol{\theta}_r^T}\right],$$

$$\left[\boldsymbol{J}_{\theta_r,\theta_s}(\boldsymbol{\theta})\right]=-E\left[\frac{\partial^2\ln f\left(\boldsymbol{y}_{1:L+K}\,|\,\boldsymbol{\theta}\right)}{\partial\boldsymbol{\theta}_r\partial\boldsymbol{\theta}_s^T}\right],$$

$$\left[\boldsymbol{J}_{\theta_s,\theta_r}(\boldsymbol{\theta})\right]=-E\left[\frac{\partial^2\ln f\left(\boldsymbol{y}_{1:L+K}\,|\,\boldsymbol{\theta}\right)}{\partial\boldsymbol{\theta}_s\partial\boldsymbol{\theta}_r^T}\right],$$

$$\left[\boldsymbol{J}_{\theta_s,\theta_s}(\boldsymbol{\theta})\right]=-E\left[\frac{\partial^2\ln f\left(\boldsymbol{y}_{1:L+K}\,|\,\boldsymbol{\theta}\right)}{\partial\boldsymbol{\theta}_s\partial\boldsymbol{\theta}_s^T}\right].$$

$f\left(\boldsymbol{y}_1,\ldots\boldsymbol{y}_L,\ldots\boldsymbol{y}_{L+K}\,|\,\boldsymbol{\theta},\boldsymbol{M}\right)$ denotes the PDF of all the data under H_1:

$$f\left(\boldsymbol{y}_{1:L+K}\,|\,\boldsymbol{\theta},\boldsymbol{M}\right)=\frac{\gamma^{-NK}}{\pi^{N(L+K)}\det^{L+K}(\boldsymbol{M})}\times\exp\left\{-\mathrm{tr}\left[\boldsymbol{M}^{-1}\left((\boldsymbol{Y}_L-\boldsymbol{v}\boldsymbol{\alpha})(\boldsymbol{Y}_L-\boldsymbol{v}\boldsymbol{\alpha})^H+\frac{1}{\gamma}\boldsymbol{S}\right)\right]\right\}\tag{2.92}$$

where $\boldsymbol{S}=\sum_{t=L+1}^{L+K}\boldsymbol{y}_t\boldsymbol{y}_t^H$.

After some calculations, the following results can be obtained

$$\frac{\partial \ln f\left(\boldsymbol{y}_{1:L+K} \mid \boldsymbol{\theta}, \boldsymbol{M}\right)}{\partial \alpha_{t,R}} = 2\mathrm{Re}\left\{\boldsymbol{v}^H \boldsymbol{M}^{-1}\left(\boldsymbol{y}_t - \alpha_t \boldsymbol{v}\right)\right\} \tag{2.93}$$

$$\frac{\partial \ln f\left(\boldsymbol{y}_{1:L+K} \mid \boldsymbol{\theta}, \boldsymbol{M}\right)}{\partial \alpha_{t,I}} = 2\mathrm{Im}\left\{\boldsymbol{v}^H \boldsymbol{M}^{-1}\left(\boldsymbol{y}_t - \alpha_t \boldsymbol{v}\right)\right\} \tag{2.94}$$

Calculating the gradient of (2.93) with respect to γ and $\boldsymbol{M}$ and taking the expectation of the results yields

$$E\left[\frac{\partial^2 \ln f\left(\boldsymbol{y}_{1:L+K} \mid \boldsymbol{\theta}\right)}{\partial \alpha_{t,R} \partial \gamma}\right] = 0 \tag{2.95}$$

$$E\left[\frac{\partial^2 \ln f\left(\boldsymbol{y}_{1:L+K} \mid \boldsymbol{\theta}\right)}{\partial \alpha_{t,R} \partial \boldsymbol{M}_{i,j}}\right] = 0 \tag{2.96}$$

$\left[\boldsymbol{J}^{-1}(\boldsymbol{\theta})\right]_{\theta_r,\theta_r}$ can be obtained by substituting Equations 2.93–2.96 into the FIM:

$$\begin{aligned}
\left[\boldsymbol{J}^{-1}(\boldsymbol{\theta})\right]_{\theta_r,\theta_r} &= \left(\boldsymbol{J}_{\theta_r,\theta_r}(\boldsymbol{\theta}) - \boldsymbol{J}_{\theta_r,\theta_s}(\boldsymbol{\theta})\boldsymbol{J}^{-1}_{\theta_s,\theta_s}(\boldsymbol{\theta})\boldsymbol{J}_{\theta_s,\theta_r}(\boldsymbol{\theta})\right)^{-1} \\
&= \boldsymbol{J}^{-1}_{\theta_r,\theta_r}(\boldsymbol{\theta}) \\
&= \frac{1}{2}\left(\boldsymbol{v}^H \boldsymbol{M}^{-1} \boldsymbol{v}\right)^{-1} \mathbf{I}_{2L}
\end{aligned} \tag{2.97}$$

From Equation 2.91, the MLE of $\boldsymbol{\theta}$ under H_0 (i.e., $\hat{\boldsymbol{\theta}}_0$) is also required to obtain the Rao test. By exploiting the persymmetric property, the PDF of all the data under H_0 can be rewritten as [27]

$$f\left(\boldsymbol{y}_{1:L+K} \mid \boldsymbol{\theta}, \boldsymbol{M}, \boldsymbol{H}_0\right) = \frac{\gamma^{-NK}}{\pi^{N(L+K)} \det^{L+K}(\boldsymbol{M})} \times \exp\left\{-\mathrm{tr}\left[\boldsymbol{M}^{-1}\left(\boldsymbol{R}_p \boldsymbol{R}_p^H + \frac{1}{\gamma}\boldsymbol{S}_p\right)\right]\right\} \tag{2.98}$$

where $\boldsymbol{S}_p = \left(\boldsymbol{S} + \boldsymbol{J}\boldsymbol{S}^*\boldsymbol{J}\right)/2$, $\boldsymbol{R}_p = \left[\boldsymbol{y}_{e1}, \ldots, \boldsymbol{y}_{eL}, \boldsymbol{y}_{o1}, \ldots, \boldsymbol{y}_{oL}\right]$, $\boldsymbol{y}_{et} = \left(\boldsymbol{y}_t + \boldsymbol{J}\boldsymbol{y}_t^*\right)/2$, $\boldsymbol{y}_{ot} = \left(\boldsymbol{y}_t - \boldsymbol{J}\boldsymbol{y}_t^*\right)/2$, $t = 1, 2, \ldots, L$.

Since the equation $\alpha_t = 0$ holds under H_0, the MLE of $\boldsymbol{\theta}_r$ under H_0 is calculated as

$$\hat{\boldsymbol{\theta}}_{r,0} = \boldsymbol{\theta}_{r,0} = \boldsymbol{0}_{2L\times 1} \tag{2.99}$$

Maximizing the PDF of the data under H_0 in Equation 2.98 over $\boldsymbol{M}$, the MLE of $\boldsymbol{M}$ under H_0 can be obtained

$$\hat{\boldsymbol{M}}_0 = \frac{1}{L+K}\left(\boldsymbol{R}_p\boldsymbol{R}_p^H + \frac{1}{\gamma}\boldsymbol{S}_p\right) \tag{2.100}$$

Plugging Equations 2.99 and 2.100 into Equation 2.98, yields

$$f\left(\boldsymbol{y}_{1:L+K} \mid \hat{\boldsymbol{M}}_0, \hat{\boldsymbol{\theta}}_{r,0}, \boldsymbol{H}_0\right) = \frac{\gamma^{-NK}}{\left(\pi^N e\right)^{(L+K)} \det^{L+K}\left(\hat{\boldsymbol{M}}_0\right)} \tag{2.101}$$

It can be seen that the MLE of γ under H_0 (i.e. $\hat{\gamma}_0$) can be obtained by minimizing $\gamma^{NK/(L+K)} \det\left(\boldsymbol{R}_p\boldsymbol{R}_p^H + \frac{1}{\gamma}\boldsymbol{S}_p\right)$ with respect to γ. According to Proposition 2 of [19], $\hat{\gamma}_0$ is the unique positive solution of the equation $\sum_{k=1}^{r_0} \frac{\lambda_k\gamma}{\lambda_k\gamma + 1} = \frac{NL}{L+K}$, where $r_0 = \min(2L, N)$ and $\lambda_k, k = 1, \ldots, r_0$ denote the nonzero eigenvalues of the matrix $\boldsymbol{S}_p^{-1/2}\boldsymbol{A}_0\boldsymbol{S}_p^{-1/2}$.

To sum up, the following results can be obtained

$$\hat{\boldsymbol{\theta}}_{s,0} = \left[\hat{\gamma}_0, \Xi\left(\hat{\boldsymbol{M}}_0\right)^T\right]^T = \left[\hat{\gamma}_0, \Xi\left(\frac{1}{L+K}\left[\boldsymbol{A}_0 + \frac{1}{\hat{\gamma}_0}\boldsymbol{S}_p\right]\right)^T\right]^T \tag{2.102}$$

where $\boldsymbol{A}_0 = \boldsymbol{R}_p\boldsymbol{R}_p^H$.

Finally, the persymmetric Rao test is obtained by substituting Equations 2.93–2.97 and Equations 2.99–2.102 into Equation 2.91 and rearranging the expression

$$\frac{\sum_{t=1}^{L}\left|v^H\left(A_0+S_p/\hat{\gamma}_0\right)^{-1}y_t\right|^2}{v^H\left(A_0+S_p/\hat{\gamma}_0\right)^{-1}v}\underset{H_0}{\overset{H_1}{\gtrless}}\eta_R' \tag{2.103}$$

where η_R' denotes the transformation of the threshold η_R.

2.4.3 The Wald Test

The real parameter Wald test is [13]

$$\hat{\theta}_{r,1}^T\left(\left[J^{-1}\left(\hat{\theta}_1\right)\right]_{\theta_r,\theta_r}\right)^{-1}\hat{\theta}_{r,1}\underset{H_0}{\overset{H_1}{\gtrless}}\eta_W \tag{2.104}$$

where $\hat{\theta}_{r,1}=\left[\hat{\alpha}_{1,R},\hat{\alpha}_{1,I},\ldots,\hat{\alpha}_{L,R},\hat{\alpha}_{L,I}\right]^T$ is the MLE of θ_r under H_1, $\hat{\theta}_1=\left[\hat{\theta}_{r,1}^T,\hat{\theta}_{s,1}^T\right]^T$ is the MLE of θ under H_1, η_W is the detection threshold.

From Equation 2.104, the MLE of θ under H_1 (i.e., $\hat{\theta}_1$) is required to obtain the Wald test. By exploiting the persymmetric property, the PDF of all the data under H_1 can be rewritten as [27]

$$f\left(y_{1:L+K}\,|\,\theta,M,H_1\right)=\frac{\gamma^{-NK}}{\pi^{N(L+K)}\det^{L+K}(M)}\times\exp\left\{-\mathrm{tr}\left[M^{-1}\left(\left(R_p-v\alpha_p\right)\left(R_p-v\alpha_p\right)^H+\frac{1}{\gamma}S_p\right)\right]\right\} \tag{2.105}$$

where $\alpha_p=\left[\alpha_{e1},\ldots,\alpha_{eL},\alpha_{o1},\ldots,\alpha_{oL}\right]$, $\alpha_{et}=\left(\alpha_t+\alpha_t^*\right)/2$, $\alpha_{ot}=\left(\alpha_t-\alpha_t^*\right)/2$, $R_p=\left[y_{e1},\ldots,y_{eL},y_{o1},\ldots,y_{oL}\right]$, $y_{et}=\left(y_t+Jy_t^*\right)/2$, $y_{ot}=\left(y_t-Jy_t^*\right)/2$, $t=1,2,\ldots,L$. Maximizing the PDF of the data under H_1 in Equation 2.105 over M, the MLE of M under H_1 can be obtained

$$\hat{M}_1=\frac{1}{L+K}\left(A_1+\frac{1}{\gamma_1}S_p\right) \tag{2.106}$$

where $A_1=\left[R_p-v\hat{\alpha}_p\right]\left[R_p-v\hat{\alpha}_p\right]^H$. Substituting Equation 2.106 into Equation 2.105, yields

$$f\left(\boldsymbol{y}_{1:L+K}|\hat{\boldsymbol{M}}_1,\boldsymbol{H}_1\right)\propto\frac{\gamma^{-NK}}{\det^{L+K}\left[\left(\boldsymbol{R}_p-\boldsymbol{v}\alpha_p\right)\left(\boldsymbol{R}_p-\boldsymbol{v}\alpha_p\right)^H+\frac{1}{\gamma}\boldsymbol{S}_p\right]} \tag{2.107}$$

From Equation 2.107, the MLE of α_p under H_1 (i.e. $\hat{\alpha}_p$) can be obtained by minimizing $\boldsymbol{T}=\det\left[\left(\boldsymbol{R}_p-\boldsymbol{v}\alpha_p\right)\left(\boldsymbol{R}_p-\boldsymbol{v}\alpha_p\right)^H+\frac{1}{\gamma}\boldsymbol{S}_p\right]$ with respect to α_p:

$$\hat{\alpha}_p=\frac{\boldsymbol{v}^H\boldsymbol{S}_p^{-1}\boldsymbol{R}_p}{\boldsymbol{v}^H\boldsymbol{S}_p^{-1}\boldsymbol{v}}=\left(\boldsymbol{v}^H\boldsymbol{S}_p^{-1}\boldsymbol{v}\right)^{-1}\cdot\left[\boldsymbol{v}^H\boldsymbol{S}_p^{-1}\boldsymbol{y}_{e1},\ldots,\boldsymbol{v}^H\boldsymbol{S}_p^{-1}\boldsymbol{y}_{eL},\boldsymbol{v}^H\boldsymbol{S}_p^{-1}\boldsymbol{y}_{o1},\ldots,\boldsymbol{v}^H\boldsymbol{S}_p^{-1}\boldsymbol{y}_{oL}\right] \tag{2.108}$$

It can be found that the MLE of γ under H_1 (i.e. $\hat{\gamma}_1$) can be obtained by minimizing $\gamma^{NK/(L+K)}\det\left[\left(\boldsymbol{R}_p-\boldsymbol{v}\breve{\alpha}_p\right)\left(\boldsymbol{R}_p-\boldsymbol{v}\breve{\alpha}_p\right)^H+\frac{1}{\gamma}\boldsymbol{S}_p\right]$ with respect to γ. From [19], $\hat{\gamma}_1$ is the unique positive solution of the equation $\sum_{k=1}^{r_1}\frac{\lambda_k\gamma}{\lambda_k\gamma+1}=\frac{NL}{L+K}$, where $r_1=\min\left(2L,N-1\right)$ and $\lambda_k,k=1,\ldots r_1$ denote the nonzero eigenvalues of the matrix $\boldsymbol{S}_p^{-1/2}\boldsymbol{A}_1\boldsymbol{S}_p^{-1/2}$. Thus,

$$\hat{\theta}_{s,1}=\left[\hat{\gamma}_1,\Xi\left(\hat{\boldsymbol{M}}_1\right)^T\right]^T=\left[\hat{\gamma}_1,\Xi\left(\frac{1}{L+K}\left[\boldsymbol{A}_1+\frac{1}{\hat{\gamma}_1}\boldsymbol{S}_p\right]\right)^T\right]^T \tag{2.109}$$

$\hat{\alpha}_p$ can also be expressed as

$$\hat{\alpha}_p=\left[\hat{\alpha}_{e1},\ldots,\hat{\alpha}_{eL},\hat{\alpha}_{o1},\ldots,\hat{\alpha}_{oL}\right]=\left[\hat{\alpha}_{1,R},\ldots,\hat{\alpha}_{L,R},j\hat{\alpha}_{1,I},\ldots,j\hat{\alpha}_{L,I}\right] \tag{2.110}$$

where $\hat{\alpha}_{t,R}$ and $\hat{\alpha}_{t,I}$ are real part and imaginary part of $\hat{\alpha}_t$. After some calculation, $\hat{\theta}_{r,1}$ can be expressed as

$$\hat{\theta}_{r,1}=[\hat{\alpha}_{1,R},\hat{\alpha}_{1,I},\ldots,\hat{\alpha}_{L,R},\hat{\alpha}_{L,I}]^T=\left[\hat{\alpha}_{e1},-j\hat{\alpha}_{o1},\ldots,\hat{\alpha}_{eL},-j\hat{\alpha}_{oL}\right]^T \tag{2.111}$$

Plugging (2.93)- (2.96) and (2.109) -(2.111) into (2.97), $\left[\boldsymbol{J}^{-1}\left(\hat{\theta}_1\right)\right]_{\theta_r,\theta_r}$ can be calculated as

$$\left[\boldsymbol{J}^{-1}\left(\hat{\theta}_1\right)\right]_{\theta_r,\theta_r} = \frac{1}{2}\left(\boldsymbol{v}^H \hat{\boldsymbol{M}}_1^{-1}\boldsymbol{v}\right)^{-1} \boldsymbol{I}_{2L\times 2L} \tag{2.112}$$

Finally, the persymmetric Wald test can be obtained by substituting Equations 2.106–2.112 into Equation 2.104 and rearranging the expression

$$\sum_{t=1}^{L}\left[\left(\frac{\boldsymbol{v}^H \boldsymbol{S}_p^{-1}\boldsymbol{y}_{et}}{\boldsymbol{v}^H \boldsymbol{S}_p^{-1}\boldsymbol{v}}\right)^2 - \left(\frac{\boldsymbol{v}^H \boldsymbol{S}_p^{-1}\boldsymbol{y}_{ot}}{\boldsymbol{v}^H \boldsymbol{S}_p^{-1}\boldsymbol{v}}\right)^2\right]\cdot\left(\boldsymbol{v}^H\left(\boldsymbol{A}_1 + \boldsymbol{S}_p/\hat{\gamma}_1\right)^{-1}\boldsymbol{v}\right) \underset{H_0}{\overset{H_1}{\gtrless}} \eta_W' \tag{2.113}$$

After some simplification, Equation 2.113 can be rewritten as

$$\hat{\gamma}_1 \sum_{t=1}^{L}\left[\left(\boldsymbol{v}^H \boldsymbol{S}_p^{-1}\boldsymbol{y}_{et}\right)^2 - \left(\boldsymbol{v}^H \boldsymbol{S}_p^{-1}\boldsymbol{y}_{ot}\right)^2\right] \Big/ \boldsymbol{v}^H \boldsymbol{S}_p^{-1}\boldsymbol{v} \underset{H_0}{\overset{H_1}{\gtrless}} \eta_W' \tag{2.114}$$

where η_W' denotes some modification of the threshold η_W.

Since only the MLEs of unknown parameters under H_0 or the MLEs of unknown parameters under H_1 are required, the newly designed Rao and Wald detectors have smaller computational complexity compared with GLRT-based detectors which require both the MLEs of unknown parameters under H_1 and the MLEs of unknown parameters under H_0.

2.4.4 Experimental Results

We evaluate the detection performance of the persymmetric Wald and persymmetric Rao (referred to as the PWald and the PRao) through simulated data and real data. The PD and the threshold are determined by resorting to Monte Carlo techniques based on 10^5 and 10^6 trials, respectively. The disturbance covariance matrix $\boldsymbol{M}$ is: $\boldsymbol{M}(i,j) = \rho^{|i-j|}, 1 \le i, j \le N$.

Here, we set $\rho = 0.9$. The SNR is defined as $\text{SNR} = \boldsymbol{v}^H \boldsymbol{M}^{-1} \boldsymbol{v} \sum_{t=1}^{L} |\alpha_t|^2$, where $\boldsymbol{v} = [1,\ldots,1]^T / \sqrt{N}$.

For comparison, we also give the detection performance of four adaptive detectors derived in a PHE: the traditional Rao test, Wald test [37], persymmetric one-step GLRT, and persymmetric two-step GLRT [30] (referred to as P1SGLRT and P2SGLRT) which also exploit the persymmetric structure of noise covariance matrix. The test statistics of the four adaptive detectors are

$$T_{Rao} = \frac{\boldsymbol{v}^H \left(\boldsymbol{R}_0 + \boldsymbol{S}/\hat{\mu}_0\right)^{-1} \boldsymbol{R}_0 \left(\boldsymbol{R}_0 + \boldsymbol{S}/\hat{\mu}_0\right)^{-1} \boldsymbol{v}}{\boldsymbol{v}^H \left(\boldsymbol{R}_0 + \boldsymbol{S}/\hat{\mu}_0\right)^{-1} \boldsymbol{v}} \underset{H_0}{\overset{H_1}{\gtrless}} \eta_{Rao} \tag{2.115}$$

$$T_{Wald} = \hat{\mu}_1 \frac{\boldsymbol{v}^H \boldsymbol{S}^{-1} \boldsymbol{Y}_L \boldsymbol{Y}_L^H \boldsymbol{S}^{-1} \boldsymbol{v}}{\boldsymbol{v}^H \boldsymbol{S}^{-1} \boldsymbol{v}} \underset{H_0}{\overset{H_1}{\gtrless}} \eta_{Wald} \tag{2.116}$$

$$T_{P1SGLRT} = \frac{\hat{\gamma}_0^{NK/(L+K)} \det\left(\boldsymbol{A}_0 + \frac{1}{\hat{\gamma}_0} \boldsymbol{S}_p\right)}{\hat{\gamma}_1^{NK/(L+K)} \det\left(\boldsymbol{A}_1 + \frac{1}{\hat{\gamma}_1} \boldsymbol{S}_p\right)} \underset{H_0}{\overset{H_1}{\gtrless}} \eta_{P1SGLRT} \tag{2.117}$$

$$T_{P2SGLRT} = \frac{\boldsymbol{v}^H \boldsymbol{S}_p^{-1} \boldsymbol{R}_p \boldsymbol{R}_p^H \boldsymbol{S}_p^{-1} \boldsymbol{v}}{\text{tr}\left(\boldsymbol{R}_p^H \boldsymbol{S}_p^{-1} \boldsymbol{R}_p\right) \boldsymbol{v}^H \boldsymbol{S}_p^{-1} \boldsymbol{v}} \underset{H_0}{\overset{H_1}{\gtrless}} \eta_{P2SGLRT} \tag{2.118}$$

where $\hat{\mu}_i, i = 0,1$ is the unique positive solution of the equation $\sum_{k=1}^{r_i} \frac{\lambda_{k,i}\mu}{\lambda_{k,i}\mu + 1} = \frac{NL}{L+K}$, where $r_0 = \min(L,N)$, $r_1 = \min(L,N-1)$.

$\lambda_{k,i}s, k = 1,\ldots r_i$ denote the nonzero eigenvalues of the matrix $\boldsymbol{S}_p^{-1/2}\boldsymbol{R}_i\boldsymbol{S}_p^{-1/2}$ where $\boldsymbol{R}_0 = \boldsymbol{Y}_L\boldsymbol{Y}_L^H$, $\boldsymbol{R}_1 = \left(\boldsymbol{Y}_L - \boldsymbol{v}\frac{\boldsymbol{v}^H\boldsymbol{S}^{-1}\boldsymbol{Y}_L}{\boldsymbol{v}^H\boldsymbol{S}^{-1}\boldsymbol{v}}\right)\left(\boldsymbol{Y}_L - \boldsymbol{v}\frac{\boldsymbol{v}^H\boldsymbol{S}^{-1}\boldsymbol{Y}_L}{\boldsymbol{v}^H\boldsymbol{S}^{-1}\boldsymbol{v}}\right)^H$.

2.4.4.1 Performance in the Presence of Matched Signals

In this part, the detection performance of the PRao and PWald is assessed in the presence of matched signals, namely, the actual steering vector $\boldsymbol{v}_m$ is aligned with the nominal one.

In Figure 2.13, the PRao and the PWald are compared with their unstructured counterparts i.e., the Rao and Wald tests. We can see that the PRao and the PWald achieved significant performance improvement over the Rao and Wald tests. The detection performance gain of the PRao with respect to the Rao test is more than 10 dB. The performance gain of the PWald with respect to the Wald test is about 6 dB for $P_d = 0.9$. When $K \leq 9$, the performance gains of the PRao and PWald with respect to the Rao test and Wald test are larger. In Figure 2.13(a), the conventional Rao

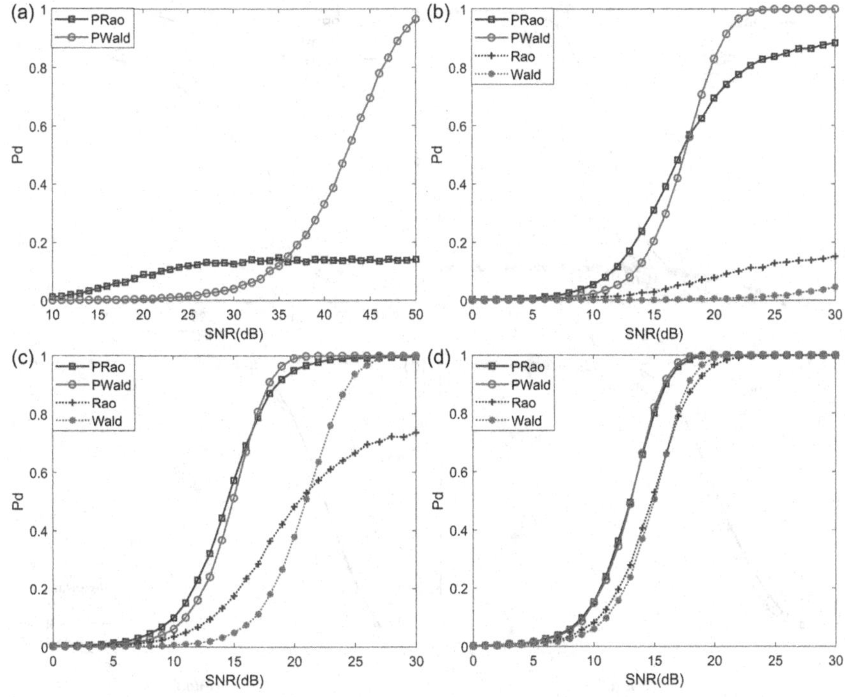

FIGURE 2.13 PD versus SNR of the PRao, the PWald, the Rao, and the Wald for $N = 9$, $L = 3$, $P_{fa} = 10^{-3}$ (a) $K = 5$; (b) $K = 9$; (c) $K = 12$; (d) $K = 18$.

and Wald detectors are not shown since the sample covariance matrix used in the Rao and Wald is singular when $K < N$. Thus, the test statistics of the Rao and Wald tests cannot be obtained. From Figure 2.13, we can also see that the PRao and Rao tests suffer detection performance degradation as the SNR becomes high. This is due to the fact that the estimated noise covariance matrix may be more contaminated by the target signal when the SNR becomes high.

We compare the detection performance of the PRao and PWald detectors with the P1SGLRT and P2SGLRT [30] which also exploit the persymmetry property in Figure 2.14. The PRao and the PWald suffer from some matched detection performance loss compared with the P1SGLRT for small data records. The performance losses of the PWald with respect to the P1SGLRT are about 5 dB and 1 dB for $P_d = 0.9$ when $K = 5$ and $K = 9$. When $K \geq 12$, the PRao and PWald achieve almost the same matched detection performance with the P1SGLRT. Meanwhile, it can be seen that the curves of probabilities of detection of the PWald are close to those of

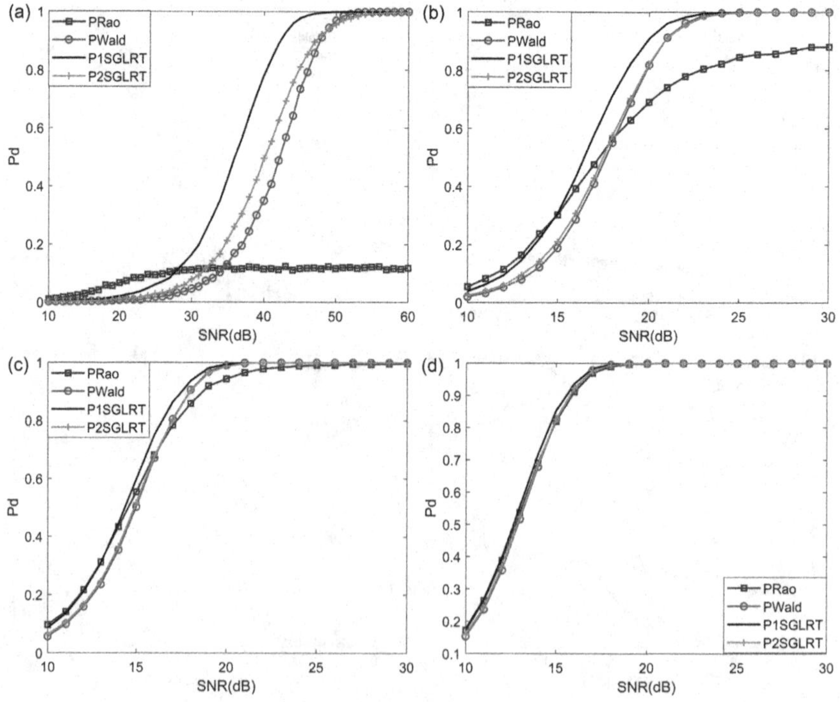

FIGURE 2.14 PD versus SNR of the PRao, the PWald, the P1SGLRT, and the P2SGLRT. (a) $K = 5$; (b) $K = 9$; (c) $K = 12$; (d) $K = 18$.

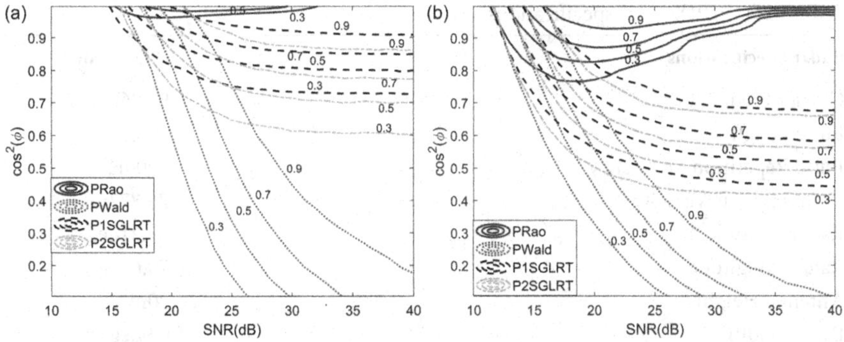

FIGURE 2.15 Contours of constant PD. (a) $K = 9$; (b) $K = 18$.

the P2SGLRT. Moreover, the PRao suffers from some detection performance degradation for high SNRs.

2.4.4.2 Performance in the Presence of Mismatched Signals

We analyze the detection performance of the PRao and PWald in the presence of mismatched signals, namely, the actual steering vector deviates from the presumed one in this part. We define the mismatch angle ϕ between $\boldsymbol{v}$ and $\boldsymbol{v}_m$ in the whitened observation space as [38] $\cos^2\phi = \frac{\left|\boldsymbol{v}^H \boldsymbol{M}^{-1}\boldsymbol{v}_m\right|^2}{\left(\boldsymbol{v}^H \boldsymbol{M}^{-1}\boldsymbol{v}\right)\left(\boldsymbol{v}_m^H \boldsymbol{M}^{-1}\boldsymbol{v}_m\right)}$. The SNR becomes $\text{SNR} = \boldsymbol{v}_m^H \boldsymbol{M}^{-1}\boldsymbol{v}_m \sum_{t=1}^{L} \left|\alpha_t\right|^2$.

We plot the contours of constant PD for the PRao, the PWald, the P1SGLRT, and the P2SGLRT for $N = 9$ and different values of K in Figure 2.15. It can be seen that compared with the P1SGLRT and the P2SGLRT, the PRao is the most selective when the signal mismatch occurs. In contrast, the PWald is the most robust to the mismatched signals.

2.4.4.3 Performance Analysis Based on the Real Data

In order to further verify the effectiveness of the proposed detectors, we investigate detection performance using real radar noise data in this subsection. The real noise data were collected by McMaster University in Canada in 1998 by irradiating Ontario Lake with the Intelligent Pixel Processing X-band (IPIX) radar in Grimsby. IPIX radar is a coherent

TABLE 2.1 IPIX radar specifications

Radar Specifications	**Value**
Carrier Frequency	9.39GHz
Pulse Length	200ns
Pulse Repetition Frequency	1000Hz
Unambiguous Velocity	7.9872m/s
Radar Latitude	43.21deg
Radar Longitude	79.6deg
Antenna Height	20m
Beam Width	0.9deg
Antenna Gain	45.7dB
Polarization Mode	HH, VV, HV, VH

radar operating in X-band with dual polarization frequency agility [39]. The IPIX radar specifications are shown in Table 2.1.

The datasets 40 (i.e. 19980205_171203_antstep.cdf) in VV polarization of McMaster IPIX radar in Grimsby are resorted to evaluate the detection performance of the detectors. There are 28 range cells, and 60000 pulses in the datasets whose range resolution is 30 m. The targets appear in the seventh range cell and the eighth range cell in the datasets. It has been verified that the datasets are compatible with the spherically invariant random vector (SIRV) model [40].

Before detection performance analysis, the statistical method proposed in [41] is used to analyze the persymmetry property of the real noise covariance matrix. The problem that decides whether the noise covariance matrix has a persymmetric structure or not is equivalent to the following binary hypothesis testing problem:

$$\begin{cases} H_0 : \boldsymbol{M} \in \mathbb{P}, \\ H_1 : \boldsymbol{M} \in \mathbb{H} - \mathbb{P}, \end{cases} \tag{2.119}$$

where $\mathbb{H}$ is the set of positive definite Hermitian matrices and $\mathbb{P}$ is the set of persymmetric matrices. Using a GLRT to solve the above binary hypothesis testing problem, yields:

$$\Gamma = \frac{\det\left[\left(\hat{\boldsymbol{M}} + \boldsymbol{J}\hat{\boldsymbol{M}}^{*}\boldsymbol{J}\right)/2\right]}{\det\left(\hat{\boldsymbol{M}}\right)} \underset{H_0}{\overset{H_1}{\gtrless}} \eta_{per} \tag{2.120}$$

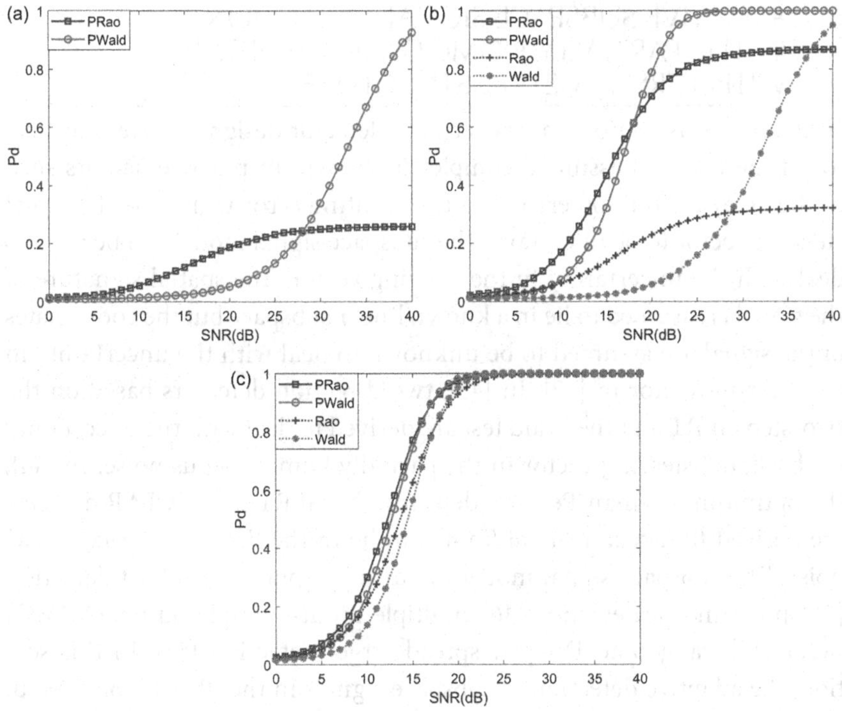

FIGURE 2.16 PD versus SNR. (a) $K = 4$; (b) $K = 7$; (c) $K = 14$.

where $\hat{\boldsymbol{M}}$ is the MLE, $\hat{\boldsymbol{M}} = \frac{1}{L+K}\sum_{t=1}^{L+K} \boldsymbol{y}_t \boldsymbol{y}_t^H$, η_{per} denotes the threshold which is selected to ensure the given value of significance level (VSL). It has been verified that the sea clutter exhibits a persymmetric covariance matrix.

In Figure 2.16, we plot the PD of the conventional Rao detector, conventional Wald detector, the PRao detector, and the PWald detector for $N = 7$ and $P_{fa} = 10^{-2}$. It can be seen that the PRao and PWald outperform the conventional Rao and Wald detectors. The detection performance gains of the PRao and PWald with respect to the Rao and Wald detectors are more than 10 dB. Both the simulated data results and the real data results demonstrate the effectiveness of the PRao and PWald.

2.5 ADAPTIVE SUBSPACE SIGNAL DETECTION IN THE PARTIALLY HOMOGENEOUS NOISE WITH PERSYMMETRIC STRUCTURE

In the previous sections, at the stage of detector design, the steering vector of the target is assumed completely known. In practice, factors such as the array calibration error and the pointing error will cause the target steering vector to be uncertain. The subspace signal model can be used to deal with the uncertainty in the steering vector. The spatial signature of the signal is assumed to lie in a known linear subspace but the coordinates of the signal are assumed to be unknown to deal with the uncertainty in the steering vector in [42]. In [43], two direction detectors based on the two-step GLRT and the Wald test are derived to deal with the uncertainty in the signal steering vector in the partially homogeneous noise. In [44], the optimum Neyman-Pearson detector, the GLRT, and a CFAR detector are derived to detect a signal known to lie in the linear subspace in CG noise. The subspace signal model can also be applied to robust detection [45], polarimetric detection [46], multiple-input-multiple-output (MIMO) radar [47], range and Doppler spread target detection [48]. In this section, the adaptive detection of subspace signals in the PHE is considered. The persymmetry property of the noise covariance matrix is exploited to enhance the matched detection performance in the limited training data case. Three CFAR detectors are derived based on the one-step GLRT, Rao, and Wald design criteria.

2.5.1 Problem Formulation

We denote the primary data by $\boldsymbol{y} \in \mathbb{C}^{N\times 1}$, which are received from N channels. The target signal lies in a full column rank subspace $\Phi \in \mathbb{C}^{N\times p}$ and is denoted by $\Phi\boldsymbol{a}$. We formulate the detection problem whether the primary data contain the target or not as the following binary hypothesis testing:

$$\begin{cases} H_0 : \begin{cases} \boldsymbol{y} = \boldsymbol{n}, \\ \boldsymbol{y}_t = \boldsymbol{n}_t, & t = 1,\ldots,K, \end{cases} \\ H_1 : \begin{cases} \boldsymbol{y} = \Phi\boldsymbol{a} + \boldsymbol{n}, \\ \boldsymbol{y}_t = \boldsymbol{n}_t, & t = 1,\ldots,K, \end{cases} \end{cases} \tag{2.121}$$

where $\boldsymbol{y}_t, t = 1,\ldots,K$ denote the training data, K is the number of the training data, $\boldsymbol{a} \in \mathbb{C}^{p\times 1}$ is the unknown coordinate, $N > p$, $\boldsymbol{n}$ and $\boldsymbol{n}_t$ denote the noise which satisfy: $\boldsymbol{n}_t \sim \mathcal{CN}_N(\boldsymbol{0}, \Sigma)$, $\boldsymbol{n} \sim \mathcal{CN}_N(\boldsymbol{0}, \tau\Sigma)$, τ is an unknown scaling factor. To model the steering vector uncertainties, p is always two or three. For polarimetric radars, p is no more than four.

2.5.2 The One-Step Generalized Likelihood Ratio Test

We assume that the covariance matrix Σ and the subspace matrix Φ are persymmetric [49, 50]: $\boldsymbol{\Sigma} = \boldsymbol{J}\boldsymbol{\Sigma}^*\boldsymbol{J}$, $\Phi = \boldsymbol{J}\Phi^*$, where $\boldsymbol{J} \in \mathbb{R}^{N\times N}$ is a permutation matrix, i.e., if $i + j = N+1$, $\boldsymbol{J}(i,j) = 1$, otherwise, $\boldsymbol{J}(i,j) = 0$.

According to the above assumption, we give the PDF for the primary data and training data under hypothesis H_i:

$$f(\boldsymbol{Y}|H_i) = \frac{1}{\pi^{N(K+1)}\tau^N \det^{(K+1)}(\Sigma)} \exp\left\{-\mathrm{tr}\left[\Sigma^{-1}\left(\frac{1}{\tau}(\boldsymbol{y} - i\Phi\boldsymbol{a})(\boldsymbol{y} - i\Phi\boldsymbol{a})^H + \boldsymbol{S}\right)\right]\right\} \tag{2.122}$$

where $\boldsymbol{Y} = [\boldsymbol{y}, \boldsymbol{y}_1, \ldots, \boldsymbol{y}_K]$, $i = 0,1$, $\boldsymbol{S} = \sum_{k=1}^{K} \boldsymbol{y}_k \boldsymbol{y}_k^H$. As derived in Appendix A in [50], the PDF can be rewritten as:

$$f(\boldsymbol{Y}|H_i) = \frac{1}{\pi^{N(K+1)}\tau^N \det^{(K+1)}(\Sigma)}$$

$$\exp\left\{-\mathrm{tr}\left[\Sigma^{-1}\left(\frac{1}{\tau}(\boldsymbol{y}_p - i\Phi\alpha)(\boldsymbol{y}_p - i\Phi\alpha)^H + \boldsymbol{S}_p\right)\right]\right\} \tag{2.123}$$

where $\boldsymbol{y}_p = [\boldsymbol{y}_e, \boldsymbol{y}_o] \in \mathbb{C}^{N\times 2}$, $\boldsymbol{y}_e = \frac{1}{2}(\boldsymbol{y} + \boldsymbol{J}\boldsymbol{y}^*)$, $\boldsymbol{y}_o = \frac{1}{2}(\boldsymbol{y} - \boldsymbol{J}\boldsymbol{y}^*)$, $\alpha = [\boldsymbol{a}_e, \boldsymbol{a}_o]$, $\boldsymbol{a}_e = \mathrm{Re}(\boldsymbol{a}), \boldsymbol{a}_o = j\mathrm{Im}(\boldsymbol{a})$, $\boldsymbol{S}_p = \frac{1}{2}(\boldsymbol{S} + \boldsymbol{J}\boldsymbol{S}^*\boldsymbol{J})$.

The one-step persymmetric subspace GLRT test statistic is given to solve the detection problem in Equation 2.121:

$$\frac{\max_{\Sigma,\alpha,\tau} f(\boldsymbol{Y}|H_1)}{\max_{\Sigma,\tau} f(\boldsymbol{Y}|H_0)} \underset{H_0}{\overset{H_1}{\gtrless}} \eta_{GLRT} \tag{2.124}$$

where η_{GLRT} denotes the threshold. It is not difficult to estimate the noise covariance matrix under the two hypotheses:

$$\hat{\Sigma}_1 = \left[\frac{1}{\tau}\left(y_p - \Phi\alpha\right)\left(y_p - \Phi\alpha\right)^H + S_p\right] \Big/ (K+1) \tag{2.125}$$

$$\hat{\Sigma}_0 = \left(\frac{1}{\tau} y_p y_p^H + S_p\right) \Big/ (K+1) \tag{2.126}$$

The test statistic Equation 2.124 is then simplified as:

$$\frac{\max\limits_{\tau} \tau^N \det^{(K+1)}\left(\frac{1}{\tau} y_p y_p^H + S_p\right)}{\max\limits_{\alpha,\tau} \tau^N \det^{(K+1)}\left[\frac{1}{\tau}\left(y_p - \Phi\alpha\right)\left(y_p - \Phi\alpha\right)^H + S_p\right]} \underset{H_0}{\overset{H_1}{\gtrless}} \eta_{GLRT} \tag{2.127}$$

To estimate α, we set $T_1 = \det\left[\frac{1}{\tau}\left(y_p - \Phi\alpha\right)\left(y_p - \Phi\alpha\right)^H + S_p\right]$ and simplify T_1 as

$$\begin{aligned} T_1 &= \det\left(S_p\right)\det\left[\frac{1}{\tau} S_p^{-1}\left(y_p - \Phi\alpha\right)\left(y_p - \Phi\alpha\right)^H + I_N\right] \\ &= \det\left(S_p\right)\det\left[\frac{1}{\tau}\left(y_p - \Phi\alpha\right)^H S_p^{-1}\left(y_p - \Phi\alpha\right) + I_2\right] \end{aligned} \tag{2.128}$$

The second equality holds since we have utilized the equation: $\det\left(B + CDE\right) = \det\left(B\right)\det\left(D\right)\det\left(D^{-1} + EB^{-1}C\right)$.

The estimation of α can be obtained by maximizing $\det\left[\frac{1}{\tau}\left(y_p - \Phi\alpha\right)^H S_p^{-1}\left(y_p - \Phi\alpha\right) + I_2\right]$ with respect to α. We take the derivative of $\det\left[\frac{1}{\tau}\left(y_p - \Phi\alpha\right)^H S_p^{-1}\left(y_p - \Phi\alpha\right) + I_2\right]$ with respect to α and get

$$\hat{\alpha} = \left(\Phi^H S_p^{-1}\Phi\right)^{-1} \Phi^H S_p^{-1} y_p \tag{2.129}$$

Substituting Equations 2.125, 2.126, and 2.129 into Equation 2.124, we obtain

$$\frac{\max_{\tau} \tau^{N/(K+1)} \det\left(\boldsymbol{I}_2 + \frac{1}{\tau} \boldsymbol{y}_p^H \boldsymbol{S}_p^{-1} \boldsymbol{y}_p \right)}{\max_{\tau} \tau^{N/(K+1)} \det\left(\boldsymbol{I}_2 + \frac{1}{\tau} \boldsymbol{y}_p^H \boldsymbol{S}_p^{-\frac{1}{2}} \boldsymbol{P}_{\Phi_{S_p}}^{\perp} \boldsymbol{S}_p^{-\frac{1}{2}} \boldsymbol{y}_p \right)} \underset{H_0}{\overset{H_1}{\gtrless}} \eta_{GLRT} \tag{2.130}$$

where $\Phi_{S_p} = \boldsymbol{S}_p^{-\frac{1}{2}}\Phi$, $\boldsymbol{P}_{\Phi_{S_p}}^{\perp} = \boldsymbol{I}_N - \boldsymbol{S}_p^{-\frac{1}{2}}\Phi\left(\Phi^H \boldsymbol{S}_p^{-1}\Phi\right)^{-1}\Phi^H \boldsymbol{S}_p^{-\frac{1}{2}}$. To get the MLE of τ, we set the ranks of $\boldsymbol{y}_p^H \boldsymbol{S}_p^{-1} \boldsymbol{y}_p$ and $\boldsymbol{y}_p^H \boldsymbol{S}_p^{-\frac{1}{2}} \boldsymbol{P}_{\Phi_{S_p}}^{\perp} \boldsymbol{S}_p^{-\frac{1}{2}} \boldsymbol{y}_p$ as r_0 and r_1, respectively. According to proposition 2 in [19], $\hat{\tau}_i$, namely, the estimation of τ under hypothesis H_i $(i = 0,1)$, is the positive solution of the equation

$$\sum_{l=1}^{r_i} \frac{\lambda_{il}}{\tau + \lambda_{il}} = \frac{N}{K+1} \tag{2.131}$$

The ranks r_0 and r_1 satisfy

$$\min\left(r_0, r_1\right) > \frac{N}{K+1} \tag{2.132}$$

where λ_{0l} and λ_{1l} are the nonzero eigenvalues of $\boldsymbol{y}_p^H \boldsymbol{S}_p^{-1} \boldsymbol{y}_p$ and $\boldsymbol{y}_p^H \boldsymbol{S}_p^{-\frac{1}{2}} \boldsymbol{P}_{\Phi_{S_p}}^{\perp} \boldsymbol{S}_p^{-\frac{1}{2}} \boldsymbol{y}_p$. Since $N > p$, it can be calculated that $r_0 = \min\left(2, N\right)$, $r_1 = \min\left(N - p, 2\right)$.

We discuss the situations to make the Equation 2.131 hold in the following.

(a) If $1 < N \leq 2$, then $r_1 = N - p$, $r_0 = N$. The equation holds when $K > 1$.

(b) If $N > 2$, then $r_0 = 2$, $r_1 = \min\left(N - p, 2\right)$. This situation can be divided into two other situations as follows. (b1) If $N - p > 2$, then $r_1 = 2$. The equation holds when $K > N/2 - 1$. (b2) If $N - p \leq 2 < N$, then $r_1 = N - p$. The equation holds when $\frac{NK}{K+1} > p$.

Finally, after some simplification, the one-step persymmetric subspace GLRT (referred to as the PSubGLRT) is

$$\frac{\hat{\tau}_0^{N/(K+1)}\det\left(\boldsymbol{I}_2+\boldsymbol{y}_p^H\boldsymbol{S}_p^{-1}\boldsymbol{y}_p/\hat{\tau}_0\right)}{\hat{\tau}_1^{N/(K+1)}\det\left(\boldsymbol{I}_2+\boldsymbol{y}_p^H\boldsymbol{S}_p^{-\frac{1}{2}}\boldsymbol{P}_{\Phi_{S_p}}^{\perp}\boldsymbol{S}_p^{-\frac{1}{2}}\boldsymbol{y}_p/\hat{\tau}_1\right)}\underset{H_0}{\overset{H_1}{\gtrless}}\eta_{PSubGLRT} \tag{2.133}$$

2.5.3 The Rao Test

We give the persymmetric subspace Rao test for the complex-valued signal

$$\left.\frac{\partial\ln f(\boldsymbol{Y}\mid H_1)}{\partial\boldsymbol{\theta}_r}\right|_{\boldsymbol{\theta}=\hat{\boldsymbol{\theta}}_0}^{T}\left[\boldsymbol{J}^{-1}\left(\hat{\boldsymbol{\theta}}_0\right)\right]_{\boldsymbol{\theta}_r,\boldsymbol{\theta}_r}\left.\frac{\partial\ln f(\boldsymbol{Y}\mid H_1)}{\partial\boldsymbol{\theta}_r^{*}}\right|_{\boldsymbol{\theta}=\hat{\boldsymbol{\theta}}_0}\underset{H_0}{\overset{H_1}{\gtrless}}\eta_{Rao} \tag{2.134}$$

where $\boldsymbol{\theta}_r=\left[\boldsymbol{a}_e^T,\boldsymbol{a}_o^T\right]^T=\mathrm{vec}(\boldsymbol{\alpha})$ is a $2p$-dimensional column vector, $\boldsymbol{\theta}_s=[\tau,\mathrm{vec}^T(\Sigma)]^T,\boldsymbol{\theta}=\left[\boldsymbol{\theta}_r^T,\boldsymbol{\theta}_s^T\right]^T$ is a $\left(2p+N^2+1\right)$-dimensional column vector, $\boldsymbol{J}(\boldsymbol{\theta})$ denotes the FIM.

We take the logarithm of the PDF for all the data under hypothesis H_1 and obtain:

$$\begin{aligned}\ln f(\boldsymbol{Y}\mid H_1)\propto&-N\ln\tau-(K+1)\ln\det(\Sigma)\\&-\mathrm{tr}\left\{\Sigma^{-1}\left[\frac{1}{\tau}(\boldsymbol{y}_p-\Phi\boldsymbol{\alpha})(\boldsymbol{y}_p-\Phi\boldsymbol{\alpha})^H+\boldsymbol{S}_p\right]\right\}\end{aligned} \tag{2.135}$$

Taking the partial derivative of Equation 2.135 with respect to $\boldsymbol{\theta}_r$, we have [51]:

$$\left.\frac{\partial\ln f(\boldsymbol{Y}\mid H_1)}{\partial\mathrm{vec}(\boldsymbol{\alpha})}\right|^{T}=\mathrm{vec}^T\left[\frac{1}{\tau}(\boldsymbol{y}_p-\Phi\boldsymbol{\alpha})^H\Sigma^{-1}\Phi\right]^T \tag{2.136}$$

$$\frac{\partial\ln f(\boldsymbol{Y}\mid H_1)}{\partial\mathrm{vec}(\boldsymbol{\alpha}^*)}=\mathrm{vec}\left[\frac{1}{\tau}\Phi^H\Sigma^{-1}(\boldsymbol{y}_p-\Phi\boldsymbol{\alpha})\right] \tag{2.137}$$

We partition $\boldsymbol{J}(\boldsymbol{\theta})$ as

$$J(\theta)=\begin{bmatrix} J_{\theta_r,\theta_r}(\theta) & J_{\theta_r,\theta_s}(\theta) \\ J_{\theta_s,\theta_r}(\theta) & J_{\theta_s,\theta_s}(\theta) \end{bmatrix} \tag{2.138}$$

According to the definition of the $J(\theta)$, we have the following results:

$$\begin{aligned}
&J_{\theta_r,\theta_r}(\theta) \\
&= E\left[\frac{\partial \ln f(\boldsymbol{Y}|H_1)}{\partial \theta_r^*}\frac{\partial \ln f(\boldsymbol{Y}|H_1)}{\partial \theta_r^T}\right] \\
&= E\left\{\text{vec}\left[\frac{1}{\tau}\Phi^H\Sigma^{-1}(\boldsymbol{y}_p-\Phi\boldsymbol{\alpha})\right]\text{vec}^T\left[\frac{1}{\tau}(\boldsymbol{y}_p-\Phi\boldsymbol{\alpha})^H\Sigma^{-1}\Phi\right]^T\right\} \\
&= E\left[\left(\boldsymbol{I}_2\otimes\frac{1}{\tau}\Phi^H\Sigma^{-1}\right)\text{vec}(\boldsymbol{y}_p-\Phi\boldsymbol{\alpha})\text{vec}^H(\boldsymbol{y}_p-\Phi\boldsymbol{\alpha})\left(\boldsymbol{I}_2\otimes\frac{1}{\tau}\Sigma^{-1}\Phi\right)\right] \\
&= \left(\boldsymbol{I}_2\otimes\frac{1}{\tau}\Phi^H\Sigma^{-1}\right)E\left[\text{vec}(\boldsymbol{y}_p-\Phi\boldsymbol{\alpha})\text{vec}^H(\boldsymbol{y}_p-\Phi\boldsymbol{\alpha})\right] \\
&\times\left(\boldsymbol{I}_2\otimes\frac{1}{\tau}\Sigma^{-1}\Phi\right) \\
&= \left(\boldsymbol{I}_2\otimes\frac{1}{\tau}\Phi^H\Sigma^{-1}\right)(\boldsymbol{I}_2\otimes\tau\Sigma)\left(\boldsymbol{I}_2\otimes\frac{1}{\tau}\Sigma^{-1}\Phi\right) \\
&= \boldsymbol{I}_2\otimes\frac{1}{\tau}\Phi^H\Sigma^{-1}\Phi
\end{aligned} \tag{2.139}$$

From Equation 2.134, it can be seen that the estimations of the unknown parameters under hypothesis H_0 are also needed to obtain the Rao test. Since $\hat{\theta}_{r,0}=\mathbf{0}$, we take the derivative of Equation 2.135 with respect to Σ and get the MLE of Σ with known τ:

$$\hat{\Sigma}_0=\left(\frac{1}{\tau}\boldsymbol{y}_p\boldsymbol{y}_p^H+\boldsymbol{S}_p\right)\Big/(K+1) \tag{2.140}$$

Substituting $\hat{\Sigma}_0$ into Equation 2.135, we know that the MLE of τ under hypothesis H_0 is equivalent to minimizing $\tau^N \det^{(K+1)}\left(\boldsymbol{y}_p\boldsymbol{y}_p^H/\tau+\boldsymbol{S}_p\right)$ with respect to τ:

$$\tau^{N}\det{}^{(K+1)}\left(\boldsymbol{y}_p\boldsymbol{y}_p^H/\tau+\boldsymbol{S}_p\right)$$
$$=\left[\det\left(\boldsymbol{S}_p\right)\tau^{N/(K+1)}\det\left(\boldsymbol{y}_p^H\boldsymbol{S}_p^{-1}\boldsymbol{y}_p/\tau+\boldsymbol{I}_2\right)\right]^{(K+1)} \tag{2.141}$$

Plugging the results into the test statistics Equation 2.134 and arranging the expressions, the persymmetric subspace Rao detector (referred to as the PSubRao) is given by:

$$\mathrm{Tr}\left[\boldsymbol{y}_p^H\hat{\Sigma}_0^{-1}\Phi\left(\Phi^H\hat{\Sigma}_0^{-1}\Phi\right)^{-1}\Phi^H\hat{\Sigma}_0^{-1}\boldsymbol{y}_p\right]\underset{H_0}{\overset{H_1}{\gtrless}}\eta_{PSubRao} \tag{2.142}$$

where $\hat{\Sigma}_0=\left(\boldsymbol{y}_p\boldsymbol{y}_p^H/\hat{\tau}_0+\boldsymbol{S}_p\right)/(K+1)$. $\hat{\tau}_0$ is the positive solution of the equation [19]

$$\sum_{l=1}^{r_0}\frac{\lambda_{0l}}{\tau+\lambda_{0l}}=\frac{N}{K+1} \tag{2.143}$$

The rank r_0 satisfies

$$r_0>\frac{N}{K+1} \tag{2.144}$$

where $r_0=\min(2,N)$, λ_{0l} is the nonzero eigenvalue of $\boldsymbol{y}_p^H\boldsymbol{S}_p^{-1}\boldsymbol{y}_p$.

We discuss the situations to make Equation 2.143 hold.

(a) If $1<N\leq 2$, then $r_0=N$. The equation holds when $K>0$.

(b) If $N>2$, then $r_0=2$. The equation holds when $K>N/2-1$.

2.5.4 The Wald Test

The persymmetric subspace Wald test for the complex-valued signal can be expressed as:

$$\hat{\boldsymbol{\theta}}_{r,1}^{H}\left(\left[\boldsymbol{J}^{-1}\left(\hat{\boldsymbol{\theta}}_{1}\right)\right]_{\theta_r,\theta_r}\right)^{-1}\hat{\boldsymbol{\theta}}_{r,1}\underset{H_0}{\overset{H_1}{\gtrless}}\eta_{Wald} \tag{2.145}$$

We take the derivative of the PDF Equation 2.123 with respect to $\boldsymbol{\alpha}$ and obtain the MLE of $\boldsymbol{\alpha}$:

$$\hat{\alpha}=\left(\Phi^{H}\Sigma^{-1}\Phi\right)^{-1}\Phi^{H}\Sigma^{-1}y_{p} \tag{2.146}$$

We substitute Equation 2.146 into Equation 2.123 and estimate Σ under hypothesis H_1:

$$\hat{\Sigma}_1=\left[\left(y_p-\Phi\hat{\alpha}\right)\left(y_p-\Phi\hat{\alpha}\right)^{H}/\tau+\boldsymbol{S}_p\right]\Big/\left(K+1\right) \tag{2.147}$$

Then, the estimation of τ under hypothesis H_1 is the positive solution of the equation [19].

$$\sum_{l=1}^{r_1}\frac{\lambda_{1l}}{\tau+\lambda_{1l}}=\frac{N}{K+1} \tag{2.148}$$

where $r_1=\min\left(N-p,2\right)$, λ_{1l} is the nonzero eigenvalue of $\boldsymbol{y}_p^{H}\boldsymbol{S}_p^{-\frac{1}{2}}\boldsymbol{P}_{\Phi_{S_p}}^{\perp}\boldsymbol{S}_p^{-\frac{1}{2}}\boldsymbol{y}_p$.

We discuss the situations to make Equation 2.148 hold.

(a) If $N-p>2$, then $r_1=2$. The equation holds when $K>N/2-1$.

(b) If $N-p\leq 2$, then $r_1=N-p$. The equation holds when $NK/\left(K+1\right)>p$.

We get the final persymmetric subspace Wald detector by plugging the above results into Equation 2.145:

$$\mathrm{Tr}\left[\boldsymbol{y}_p^{H}\hat{\Sigma}_1^{-1}\Phi\left(\Phi^{H}\hat{\Sigma}_1^{-1}\Phi\right)^{-1}\Phi^{H}\hat{\Sigma}_1^{-1}\boldsymbol{y}_p\right]\Big/\hat{\tau}_1\underset{H_0}{\overset{H_1}{\gtrless}}\eta_{PSubWald} \tag{2.149}$$

where $\hat{\Sigma}_1 = \left[\frac{1}{\hat{\tau}_1}(\boldsymbol{y}_p - \Phi\hat{\boldsymbol{\alpha}})(\boldsymbol{y}_p - \Phi\hat{\boldsymbol{\alpha}})^H + \boldsymbol{S}_p\right] \Big/ (K+1)$.

We simplify $\hat{\Sigma}_1^{-1}$ by using the woodbury formula [52] and ignore the constant:

$$\hat{\Sigma}_1^{-1} \propto \boldsymbol{S}_p^{-1} - \frac{1}{\hat{\tau}_1}\boldsymbol{S}_p^{-1}\boldsymbol{B}_p\left(\boldsymbol{I}_2 + \frac{1}{\hat{\tau}_1}\boldsymbol{y}_p^H\boldsymbol{S}_p^{-1}\boldsymbol{B}_p\right)^{-1}\boldsymbol{B}_p^H\boldsymbol{S}_p^{-1} \tag{2.150}$$

where $\boldsymbol{B}_p = \left[\boldsymbol{y}_p - \Phi\left(\Phi^H\boldsymbol{S}_p^{-1}\Phi\right)^{-1}\Phi^H\boldsymbol{S}_p^{-1}\boldsymbol{y}_p\right]$, $\propto$ denotes "proportional to". After some calculation, we can get the following results:

$$\Phi^H\boldsymbol{S}_p^{-1}\boldsymbol{B}_p = \Phi^H\boldsymbol{S}_p^{-1}\boldsymbol{y}_p - \Phi^H\boldsymbol{S}_p^{-1}\Phi\left(\Phi^H\boldsymbol{S}_p^{-1}\Phi\right)^{-1}\Phi^H\boldsymbol{S}_p^{-1}\boldsymbol{y}_p = \boldsymbol{0} \tag{2.151}$$

$$\begin{aligned}\Phi^H\hat{\Sigma}_1^{-1}\Phi &\propto \Phi^H\boldsymbol{S}_p^{-1}\Phi - \frac{1}{\hat{\tau}_1}\Phi^H\boldsymbol{S}_p^{-1}\boldsymbol{B}_p\left(\boldsymbol{I}_2 + \frac{1}{\hat{\tau}_1}\boldsymbol{y}_p^H\boldsymbol{S}_p^{-1}\boldsymbol{B}_p\right)^{-1}\boldsymbol{B}_p^H\boldsymbol{S}_p^{-1}\Phi \\ &\propto \Phi^H\boldsymbol{S}_p^{-1}\Phi\end{aligned} \tag{2.152}$$

$$\begin{aligned}\Phi^H\hat{\Sigma}_1^{-1}\boldsymbol{y}_p &\propto \Phi^H\boldsymbol{S}_p^{-1}\boldsymbol{y}_p - \frac{1}{\hat{\tau}_1}\Phi^H\boldsymbol{S}_p^{-1}\boldsymbol{B}_p\left(\boldsymbol{I}_2 + \frac{1}{\hat{\tau}_1}\boldsymbol{y}_p^H\boldsymbol{S}_p^{-1}\boldsymbol{B}_p\right)^{-1}\boldsymbol{B}_p^H\boldsymbol{S}_p^{-1}\boldsymbol{y}_p \\ &\propto \Phi^H\boldsymbol{S}_p^{-1}\boldsymbol{y}_p\end{aligned} \tag{2.153}$$

Consequently, the persymmetric subspace Wald detector (referred to as the PSubWald) is

$$\mathrm{Tr}\left[\boldsymbol{y}_p^H\boldsymbol{S}_p^{-1}\Phi\left(\Phi^H\boldsymbol{S}_p^{-1}\Phi\right)^{-1}\Phi^H\boldsymbol{S}_p^{-1}\boldsymbol{y}_p\right] \Big/ \hat{\tau}_1 \underset{H_0}{\overset{H_1}{\gtrless}} \eta_{PSubWald} \tag{2.154}$$

2.5.5 CFAR Property Analysis

We introduce a unitary matrix $\boldsymbol{T}=\frac{1}{2}\left[\left(\boldsymbol{I}_N+\boldsymbol{J}\right)+j\left(\boldsymbol{I}_N-\boldsymbol{J}\right)\right]$ to transform $\boldsymbol{y}_e$, $\boldsymbol{y}_o$, $\boldsymbol{S}_p$, and Φ to real matrices:

$$\boldsymbol{y}_{er}=\boldsymbol{T}\boldsymbol{y}_e=\frac{1}{2}\left[\left(\boldsymbol{I}_N+\boldsymbol{J}\right)\mathrm{Re}\left(\boldsymbol{y}\right)-\left(\boldsymbol{I}_N-\boldsymbol{J}\right)\mathrm{Im}\left(\boldsymbol{y}\right)\right] \tag{2.155}$$

$$\boldsymbol{y}_{or}=-j\boldsymbol{T}\boldsymbol{y}_o=\frac{1}{2}\left[\left(\boldsymbol{I}_N-\boldsymbol{J}\right)\mathrm{Re}\left(\boldsymbol{y}\right)+\left(\boldsymbol{I}_N+\boldsymbol{J}\right)\mathrm{Im}\left(\boldsymbol{y}\right)\right] \tag{2.156}$$

$$\tilde{\boldsymbol{S}}_p=\boldsymbol{T}\boldsymbol{S}_p\boldsymbol{T}^H=\mathrm{Re}\left(\boldsymbol{S}_p\right)+\boldsymbol{J}\,\mathrm{Im}\left(\boldsymbol{S}_p\right) \tag{2.157}$$

$$\tilde{\Phi}=\boldsymbol{T}\Phi=\mathrm{Re}\left(\Phi\right)-\mathrm{Im}\left(\Phi\right) \tag{2.158}$$

Since $\boldsymbol{y}$ is circular, we have $E\left(\boldsymbol{y}\boldsymbol{y}^T\right)=\boldsymbol{0}$, $E\left(\boldsymbol{y}_e\boldsymbol{y}_e^H\right)=E\left(\boldsymbol{y}_o\boldsymbol{y}_o^H\right)=\Sigma/2$, and $E\left(\boldsymbol{y}_{er}\boldsymbol{y}_{er}^H\right)=E\left(\boldsymbol{y}_{or}\boldsymbol{y}_{or}^H\right)=\boldsymbol{T}\Sigma\boldsymbol{T}^H/2=\Sigma_r$. We set $\tilde{\boldsymbol{y}}_p=\left[\boldsymbol{y}_{er},\boldsymbol{y}_{or}\right]\in R^{N\times 2}$. Then, we get $\tilde{\boldsymbol{y}}_p=\boldsymbol{T}\boldsymbol{y}_p\begin{bmatrix}1 & 0\\ 0 & -j\end{bmatrix}=\boldsymbol{T}\boldsymbol{y}_p\boldsymbol{V}$, where $\boldsymbol{V}$ is a unitary matrix.

Let $\boldsymbol{y}_p'=\Sigma_r^{-\frac{1}{2}}\tilde{\boldsymbol{y}}_p/\sqrt{\tau}$, $\Phi'=\Sigma_r^{-\frac{1}{2}}\tilde{\Phi}$, $\boldsymbol{S}_p'=\Sigma_r^{-\frac{1}{2}}\tilde{\boldsymbol{S}}_p\Sigma_r^{-\frac{1}{2}}$. An orthogonal matrix $\boldsymbol{Q}=\left[\boldsymbol{Q}_1,\boldsymbol{Q}_2\right]$ exists such that $\boldsymbol{Q}_1=\Phi'\left(\Phi'^H\Phi'\right)^{-\frac{1}{2}}$, $\boldsymbol{Q}_2^T\boldsymbol{Q}_1=\boldsymbol{0}_{(N-p)\times p}$, $\boldsymbol{Q}_2^T\boldsymbol{Q}_2=\boldsymbol{0}_{(N-p)\times(N-p)}$. We define $\breve{\Phi}=\boldsymbol{Q}^T\Phi'$, $\breve{\boldsymbol{y}}_p=\boldsymbol{Q}^T\boldsymbol{y}_p'=\begin{bmatrix}\breve{\boldsymbol{y}}_{p1}\\ \breve{\boldsymbol{y}}_{p2}\end{bmatrix}=\begin{bmatrix}\boldsymbol{Q}_1^T\boldsymbol{y}_p'\\ \boldsymbol{Q}_2^T\boldsymbol{y}_p'\end{bmatrix}$, $\breve{\boldsymbol{S}}_p=\begin{bmatrix}\breve{\boldsymbol{S}}_{p11} & \breve{\boldsymbol{S}}_{p12}\\ \breve{\boldsymbol{S}}_{p21} & \breve{\boldsymbol{S}}_{p22}\end{bmatrix}=\boldsymbol{Q}^T\boldsymbol{S}_p'\boldsymbol{Q}$.

We use the newly defined matrices to represent the original ones and utilize the properties of the unitary matrix and orthogonal matrix to simplify the expressions related to the proposed detectors. We have the following equations:

$$
\begin{aligned}
&\boldsymbol{y}_p^H \boldsymbol{S}_p^{-1} \boldsymbol{y}_p - \boldsymbol{y}_p^H \boldsymbol{S}_p^{-1} \Phi \left(\Phi^H \boldsymbol{S}_p^{-1} \Phi \right)^{-1} \Phi^H \boldsymbol{S}_p^{-1} \boldsymbol{y}_p \\
&= \boldsymbol{V} \tilde{\boldsymbol{y}}_p^H \tilde{\boldsymbol{S}}_p^{-1} \tilde{\boldsymbol{y}}_p \boldsymbol{V}^H - \boldsymbol{V} \tilde{\boldsymbol{y}}_p^H \tilde{\boldsymbol{S}}_p^{-1} \tilde{\Phi} \left(\tilde{\Phi}^H \tilde{\boldsymbol{S}}_p^{-1} \tilde{\Phi} \right)^{-1} \tilde{\Phi}^H \tilde{\boldsymbol{S}}_p^{-1} \tilde{\boldsymbol{y}}_p \boldsymbol{V}^H \\
&= \boldsymbol{V} \tilde{\boldsymbol{y}}_p^H \left[\tilde{\boldsymbol{S}}_p^{-1} - \tilde{\boldsymbol{S}}_p^{-1} \tilde{\Phi} \left(\tilde{\Phi}^H \tilde{\boldsymbol{S}}_p^{-1} \tilde{\Phi} \right)^{-1} \tilde{\Phi}^H \tilde{\boldsymbol{S}}_p^{-1} \right] \tilde{\boldsymbol{y}}_p \boldsymbol{V}^H \\
&= \boldsymbol{V} \tilde{\boldsymbol{y}}_p^H \tilde{\boldsymbol{G}} \tilde{\boldsymbol{y}}_p \boldsymbol{V}^H = \tau \boldsymbol{V} \boldsymbol{y}_p'^H \boldsymbol{G}' \boldsymbol{y}_p' \boldsymbol{V}^H = \tau \boldsymbol{V} \breve{\boldsymbol{y}}_p^H \breve{\boldsymbol{G}} \breve{\boldsymbol{y}}_p \boldsymbol{V}^H \\
&= \tau \boldsymbol{V} \breve{\boldsymbol{y}}_{p2}^H \breve{\boldsymbol{G}}_{22} \breve{\boldsymbol{y}}_{p2} \boldsymbol{V}^H
\end{aligned}
\tag{2.159}
$$

where $\tilde{\boldsymbol{G}} = \left[\tilde{\boldsymbol{S}}_p^{-1} - \tilde{\boldsymbol{S}}_p^{-1} \tilde{\Phi} \left(\tilde{\Phi}^H \tilde{\boldsymbol{S}}_p^{-1} \tilde{\Phi} \right)^{-1} \tilde{\Phi}^H \tilde{\boldsymbol{S}}_p^{-1} \right]$,

$\boldsymbol{G}' = \left[\boldsymbol{S}_p'^{-1} - \boldsymbol{S}_p'^{-1} \tilde{\Phi}' \left(\tilde{\Phi}'^H \boldsymbol{S}_p'^{-1} \tilde{\Phi}' \right)^{-1} \tilde{\Phi}'^H \boldsymbol{S}_p'^{-1} \right]$,

$\breve{\boldsymbol{G}} = \left[\breve{\boldsymbol{S}}_p^{-1} - \breve{\boldsymbol{S}}_p^{-1} \breve{\Phi} \left(\breve{\Phi}^H \breve{\boldsymbol{S}}_p^{-1} \breve{\Phi} \right)^{-1} \breve{\Phi}^H \breve{\boldsymbol{S}}_p^{-1} \right] = \boldsymbol{Q}^T \boldsymbol{G}' \boldsymbol{Q}$. The last two equalities hold since $\breve{\boldsymbol{y}}_{p2} = \boldsymbol{Q}_2^T \boldsymbol{y}_p'$, $\boldsymbol{Q}_1^T \boldsymbol{G}' = \boldsymbol{0}$, $\breve{\boldsymbol{G}} = \boldsymbol{Q}^T \boldsymbol{G}' \boldsymbol{Q} = \begin{bmatrix} \boldsymbol{0}_{p \times p} & \boldsymbol{0}_{p \times (N-p)} \\ \boldsymbol{0}_{(N-p) \times p} & \breve{\boldsymbol{G}}_{22} \end{bmatrix}$,

$\breve{\boldsymbol{G}}_{22} = \boldsymbol{Q}_2^T \boldsymbol{G}' \boldsymbol{Q}_2$.

We give the inverse of $\breve{\boldsymbol{S}}_p$ and utilize the partitioned matrix inversion theorem to rewrite $\breve{\boldsymbol{G}}_{22}$ as

$$
\begin{aligned}
\breve{\boldsymbol{S}}_p^{-1} &= \begin{bmatrix} \breve{\boldsymbol{S}}'_{p11} & \breve{\boldsymbol{S}}'_{p12} \\ \breve{\boldsymbol{S}}'_{p21} & \breve{\boldsymbol{S}}'_{p22} \end{bmatrix} \\
&= \begin{bmatrix} \boldsymbol{Q}_1^T \boldsymbol{S}_p'^{-1} \boldsymbol{Q}_1 & \boldsymbol{Q}_1^T \boldsymbol{S}_p'^{-1} \boldsymbol{Q}_2 \\ \boldsymbol{Q}_2^T \boldsymbol{S}_p'^{-1} \boldsymbol{Q}_1 & \boldsymbol{Q}_2^T \boldsymbol{S}_p'^{-1} \boldsymbol{Q}_2 \end{bmatrix}
\end{aligned}
\tag{2.160}
$$

$$
\begin{aligned}
\breve{\boldsymbol{G}}_{22} &= \left[\boldsymbol{Q}_2^T \boldsymbol{S}_p'^{-1} \boldsymbol{Q}_2 - \boldsymbol{Q}_2^T \boldsymbol{S}_p'^{-1} \tilde{\Phi}' \left(\tilde{\Phi}'^H \boldsymbol{S}_p'^{-1} \tilde{\Phi}' \right)^{-1} \tilde{\Phi}'^H \boldsymbol{S}_p'^{-1} \boldsymbol{Q}_2 \right] \\
&= \breve{\boldsymbol{S}}'_{p22} - \breve{\boldsymbol{S}}'_{p21} \breve{\boldsymbol{S}}'^{-1}_{p11} \breve{\boldsymbol{S}}'_{p12} = \breve{\boldsymbol{S}}_{p22}^{-1}
\end{aligned}
\tag{2.161}
$$

The expression $\boldsymbol{y}_p^H \boldsymbol{S}_p^{-1} \Phi \left(\Phi^H \boldsymbol{S}_p^{-1} \Phi \right)^{-1} \Phi^H \boldsymbol{S}_p^{-1} \boldsymbol{y}_p$ can then be simplified as

$$
\begin{aligned}
& y_p^H S_p^{-1}\Phi\left(\Phi^H S_p^{-1}\Phi\right)^{-1}\Phi^H S_p^{-1} y_p \\
&= V\tilde{y}_p^H \tilde{S}_p^{-1}\tilde{\Phi}\left(\tilde{\Phi}^H \tilde{S}_p^{-1}\tilde{\Phi}\right)^{-1}\tilde{\Phi}^H \tilde{S}_p^{-1}\tilde{y}_p V^H \\
&= \tau V\tilde{y}_p'^H S_p'^{-1} Q_1\left(Q_1^H \tilde{S}_p'^{-1} Q_1\right)^{-1} Q_1^H S_p'^{-1}\tilde{y}_p' V^H \\
&= \tau V\breve{y}_p^H \breve{S}_p^{-1}\breve{Q}_1\left(\breve{Q}_1^H \breve{S}_p^{-1}\breve{Q}_1\right)^{-1}\breve{Q}_1^H \breve{S}_p^{-1}\breve{y}_p V^H \\
&= \tau V\left(\breve{y}_{p1}^H \breve{S}'_{p11} + \breve{y}_{p2}^H \breve{S}'_{p21}\right)\breve{S}'^{-1}_{p11}\left(\breve{y}_{p1}^H \breve{S}'_{p11} + \breve{y}_{p2}^H \breve{S}'_{p21}\right)^H V^H \\
&= \tau V\left(\breve{y}_{p1} - \breve{S}_{p12}\breve{S}_{p22}^{-1}\breve{y}_{p2}\right)^H \breve{S}'_{p11}\left(\breve{y}_{p1} - \breve{S}_{p12}\breve{S}_{p22}^{-1}\breve{y}_{p2}\right)V^H
\end{aligned}
\tag{2.162}
$$

where $\breve{Q}_1 = Q^T Q_1 = \begin{bmatrix} I_p \\ 0_{(N-p)\times p} \end{bmatrix}$. The last equality holds since $\breve{S}'_{p11} = \left(\breve{S}_{p11} - \breve{S}_{p12}\breve{S}_{p22}^{-1}\breve{S}_{p21}\right)^{-1}$, $\breve{S}'_{p21} = -\breve{S}_{p22}^{-1}\breve{S}_{p21}\breve{S}'_{p11}$.

The expression $y_p^H \hat{\Sigma}_0^{-1}\Phi\left(\Phi^H \hat{\Sigma}_0^{-1}\Phi\right)^{-1}\Phi^H \hat{\Sigma}_0^{-1} y_p$ can be simplified in a similar manner. The only difference is that $\hat{\Sigma}_0^{-1}$ is calculated instead of S_p^{-1} in this expression. After some calculation, the expression $y_p^H \hat{\Sigma}_0^{-1}\Phi\left(\Phi^H \hat{\Sigma}_0^{-1}\Phi\right)^{-1}\Phi^H \hat{\Sigma}_0^{-1} y_p$ reduces to:

$$
\begin{aligned}
& y_p^H \hat{\Sigma}_0^{-1}\Phi\left(\Phi^H \hat{\Sigma}_0^{-1}\Phi\right)^{-1}\Phi^H \hat{\Sigma}_0^{-1} y_p \\
&= \tau V\breve{y}_p^H \breve{\Sigma}_0^{-1}\breve{Q}_1\left(\breve{Q}_1^H \breve{\Sigma}_0^{-1}\breve{Q}_1\right)^{-1}\breve{Q}_1^H \breve{\Sigma}_0^{-1}\breve{y}_p V^H
\end{aligned}
\tag{2.163}
$$

where $\breve{\Sigma}_0 \propto \frac{1}{\breve{\tau}_0} Q^T \Sigma_r^{-\frac{1}{2}} T\left(y_p y_p^H + S_p\right)T^H \Sigma_r^{-\frac{1}{2}} Q \propto \frac{1}{\breve{\tau}_0}\left(\tau\breve{y}_p\breve{y}_p^H + \breve{S}_p\right)$.

The expression $y_p^H S_p^{-1} y_p$ can be simplified as

$$
\begin{aligned}
& y_p^H S_p^{-1} y_p = V\tilde{y}_p^H \tilde{S}_p^{-1}\tilde{y}_p V^H \\
&= \tau V\tilde{y}_p'^H \tilde{S}_p'^{-1}\tilde{y}_p' V^H = \tau V\breve{y}_p^H \breve{S}_p^{-1}\breve{y}_p V^H \\
&= \tau V\begin{bmatrix} \breve{y}_{p1}^H & \breve{y}_{p2}^H \end{bmatrix}\begin{bmatrix} \breve{S}'_{p11} & \breve{S}'_{p12} \\ \breve{S}'_{p21} & \breve{S}'_{p22} \end{bmatrix}\begin{bmatrix} \breve{y}_{p1} \\ \breve{y}_{p2} \end{bmatrix} V^H \\
&= \tau V\left(\breve{y}_{p1} - \breve{S}_{p12}\breve{S}_{p22}^{-1}\breve{y}_{p2}\right)^H \breve{S}'_{p11}\left(\breve{y}_{p1} - \breve{S}_{p12}\breve{S}_{p22}^{-1}\breve{y}_{p2}\right)V^H \\
&+ \tau V\breve{y}_{p2}^H \breve{S}_{p22}^{-1}\breve{y}_{p2} V^H
\end{aligned}
\tag{2.164}
$$

It has been derived in Appendix C in [50] that $\tilde{S}_p \sim W_N(2K, \Sigma_r)$, $\tilde{y}_p \sim N(\mathbf{0}, \Sigma_r \otimes I_2)$. After some calculation, we have $\breve{S}_p \sim W_N(2K, I_r)$, $\breve{y}_p \sim N(\mathbf{0}, I_r \otimes I_2)$ under hypothesis H_0. $\breve{y}_p$ and $\breve{S}_p$ are both independent of Σ.

Since $\breve{\tau}_0$ and $\breve{\tau}_1$ are the solutions of the equations related to the nonzero eigenvalues of $y_p^H S_p^{-1} y_p$ and $y_p^H S_p^{-\frac{1}{2}} P_{\Phi_{S_p}}^{\perp} S_p^{-\frac{1}{2}} y_p$, the test statistics of the derived detectors are functions of expressions $y_p^H S_p^{-1} y_p$, $y_p^H S_p^{-1}\Phi\left(\Phi^H S_p^{-1}\Phi\right)^{-1}\Phi^H S_p^{-1} y_p$, $y_p^H S_p^{-\frac{1}{2}} P_{\Phi_{S_p}}^{\perp} S_p^{-\frac{1}{2}} y_p$, and $y_p^H \breve{\Sigma}_0^{-1}\Phi\left(\Phi^H \breve{\Sigma}_0^{-1}\Phi\right)^{-1}\Phi^H \breve{\Sigma}_0^{-1} y_p$. The independence of these expressions with respect to Σ indicates that the newly proposed persymmetric subspace detectors are CFAR with respect to noise covariance matrix.

2.5.6 Experimental Results

2.5.6.1 Performance Analysis Based on Simulated Data

To obtain the detection threshold and PD, 10^6 independent Monte Carlo simulation trials are conducted. We set $N = 12$, $p = 3$, $\tau = 2$, $P_{fa} = 10^{-3}$, $\Phi = \left[v(f_{d1}), \ldots, v(f_{dp})\right]$, $v(f_{di}) = \left[e^{-j2\pi f_{di}\frac{(N-1)}{2}}, \ldots, e^{-j2\pi f_{di}\frac{1}{2}}, e^{j2\pi f_{di}\frac{1}{2}}, \ldots, e^{j2\pi f_{di}\frac{(N-1)}{2}}\right]^T$, $i = 1, \ldots, p$, unless otherwise specified. The noise covariance matrix is

$$\Sigma_{i,j} = \rho^{|i-j|}, 1 \le i, j \le N \tag{2.165}$$

where $\rho = 0.9$. We define the SNR as:

$$\text{SNR} = \text{trace}\left(a^H \Phi^H \Sigma^{-1} \Phi a\right) \tag{2.166}$$

In Figure 2.17, the PFAs of the PSubGLRT, PSubRao, and PSubWald are plotted versus correlation coefficients ρ. It can be seen that the PFA remains unchanged for various correlation coefficients ρ. Thus, the CFAR property of the proposed detectors is demonstrated.

In Figure 2.18, the detection performance of the PSubGLRT, PSubRao, and PSubWald is compared with the PASD [53], the persymmetric

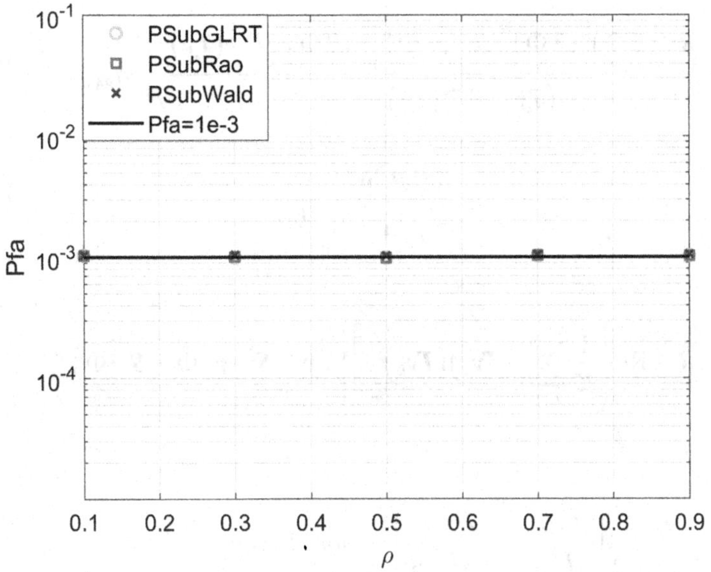

FIGURE 2.17 PFA versus ρ.

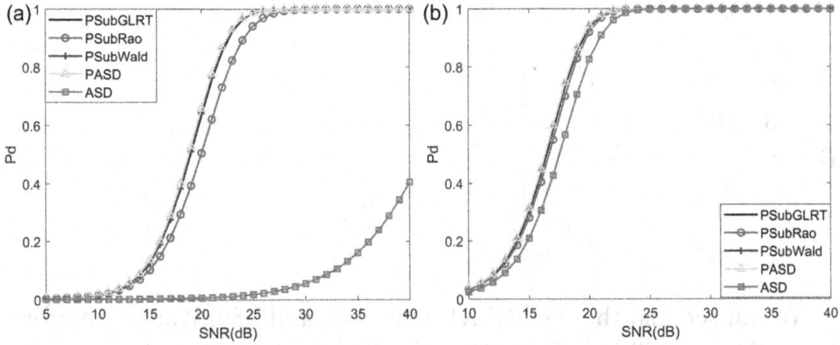

FIGURE 2.18 Detection performance of the detectors for $N=12, p=3$ and various values of K. (a) $K=12$; (b) $K=24$.

adaptive coherence estimator (PACE) [54], and the adaptive subspace detector (ASD) [48] when no signal mismatch occurs. When $p=1$, the PASD is exactly the PACE. For clarification, the test statistics of the PASD and ASD are given by:

$$\frac{(\boldsymbol{T}\boldsymbol{y})^H \hat{\boldsymbol{R}}^{-1}(\boldsymbol{T}\Phi)\left[(\boldsymbol{T}\Phi)^H \hat{\boldsymbol{R}}^{-1}(\boldsymbol{T}\Phi)\right]^{-1}(\boldsymbol{T}\Phi)^H \hat{\boldsymbol{R}}^{-1}(\boldsymbol{T}\boldsymbol{y})}{(\boldsymbol{T}\boldsymbol{y})^H \hat{\boldsymbol{R}}^{-1}(\boldsymbol{T}\boldsymbol{y})} \underset{H_0}{\overset{H_1}{\gtrless}} \eta_{PASD} \tag{2.167}$$

$$\frac{\dot{\boldsymbol{y}}^H \boldsymbol{P}_{\dot{\Phi}} \dot{\boldsymbol{y}}}{\dot{\boldsymbol{y}}^H \dot{\boldsymbol{y}}} \underset{H_0}{\overset{H_1}{\gtrless}} \eta_{ASD} \tag{2.168}$$

where $\hat{\boldsymbol{R}} = \mathrm{Re}\left\{\frac{1}{K}\sum_{k=1}^{K}\left[(\boldsymbol{T}\boldsymbol{y}_k)(\boldsymbol{T}\boldsymbol{y}_k)^H\right]\right\}$, $\dot{\boldsymbol{y}} = \boldsymbol{S}^{-\frac{1}{2}}\boldsymbol{y}$, $\dot{\Phi} = \boldsymbol{S}^{-\frac{1}{2}}\Phi$,

$$\boldsymbol{T} = \begin{cases} \frac{1}{\sqrt{2}}\begin{pmatrix} \boldsymbol{I}_{N/2} & \boldsymbol{D}_{N/2} \\ j\boldsymbol{I}_{N/2} & -j\boldsymbol{D}_{N/2} \end{pmatrix} & \text{for } N \text{ even} \\ \frac{1}{\sqrt{2}}\begin{pmatrix} \boldsymbol{I}_{(N-1)/2} & 0 & \boldsymbol{D}_{(N-1)/2} \\ 0 & \sqrt{2} & 0 \\ j\boldsymbol{I}_{(N-1)/2} & 0 & -j\boldsymbol{D}_{(N-1)/2} \end{pmatrix} & \text{for } N \text{ odd} \end{cases},$$

$$\boldsymbol{D} = \begin{bmatrix} 0 & 0 & \cdots & 0 & 1 \\ 0 & 0 & \cdots & 1 & 0 \\ \vdots & \vdots & \vdots & \vdots & \vdots \\ 0 & 1 & \cdots & 0 & 0 \\ 1 & 0 & \cdots & 0 & 0 \end{bmatrix}.$$

We can see that the PSubGLRT, PSubRao, and PSubWald outperform the ASD, especially in the training-limited case. In Figure 2.18 (a), the PSubGLRT, PSubWald, and PASD have similar detection performance. The performance improvement of the PSubGLRT, PSubRao, and PSubWald with respect to the ASD is more than 10 dB for $K = 12$. This is due to the fact that the exploitation of the noise covariance matrix property helps reduce the number of training data required to obtain satisfactory detection performance. Meanwhile, the PSubRao suffers from about 1.5 dB performance loss at $P_d = 0.9$ compared to the PSubGLRT and PSubWald.

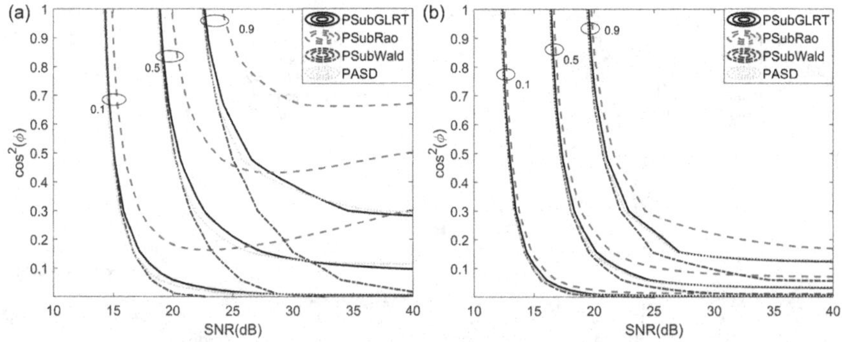

FIGURE 2.19 Contours of const PD for the PSubGLRT, PSubRao, PSubWald, and PASD. (a) $K = 12$; (b) $K = 24$.

When the number of training data increases to $K = 2N$, the probability curves of all the detectors are almost the same.

To see the detection performance of the PSubGLRT, PSubRao, and PSubWald when the signal mismatch occurs, we plot the contours of const PD in Figure 2.19. We define the mismatched angle ϕ as [50] $\cos^2\phi = \frac{\left|\boldsymbol{v}^H \Sigma^{-1} \boldsymbol{v}_m\right|^2}{\left(\boldsymbol{v}^H \Sigma^{-1} \boldsymbol{v}\right)\left(\boldsymbol{v}_m^H \Sigma^{-1} \boldsymbol{v}_m\right)}$, where $\boldsymbol{v}$ and $\boldsymbol{v}_m$ denote the nominal steering vector and the true steering vector. It can be seen that the PSubWald is the most robust to the mismatched signals. The PSubRao is more selective than the other three persymmetric subspace detectors. The PASD is a bit more robust than the PSubGLRT under the condition of low SNR, while the PSubGLRT has better robustness when SNR becomes high.

To see the influence of p on the detection performance, the PDs of the proposed detectors are plotted versus SNR for $p = 2$ and $p = 1$ in matched and mismatched signal cases in Figure 2.20. When $p = 1$, the proposed persymmetric subspace detectors reduce to the persymmetric rank-one signal detectors.

It can be seen that the PSubGLRT, PSubWald, and PSubRao achieve significant performance gains compared with the persymmetric rank-one signal detectors in the mismatched signal case. Meanwhile, the performance losses of the PSubGLRT, PSubWald, and PSubRao with respect to the persymmetric rank-one signal detectors are about 2.5 dB in the

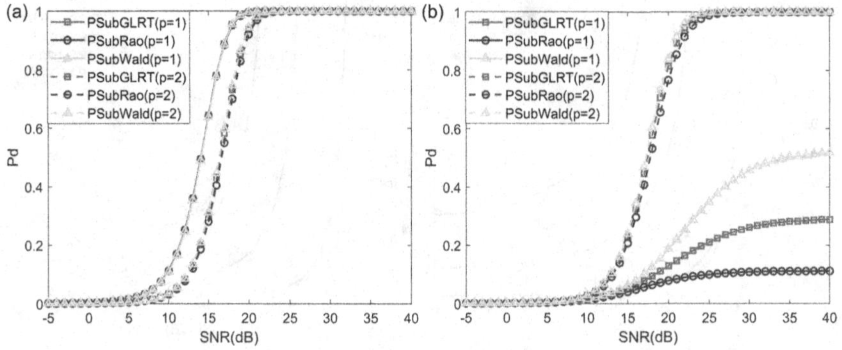

FIGURE 2.20 PD versus SNR for $N=10, K=10$. (a) $\cos^2\phi=1$; (b) $\cos^2\phi=0.4795$.

matched signal case. Thus, the subspace signal model can help improve the robustness.

2.5.6.2 *Performance Analysis Based on Real Data*

The datasets 40 (i.e. 19980205_171203_antstep.cdf) in HH polarization of McMaster IPIX radar in Grimsby are used to evaluate the detection performance of the PSubGLRT, PSubWald, and PSubRao further. The datasets have been demonstrated to satisfy the SIRV model [40]. In Figure 2.21, the detection performance of the PSubGLRT, PSubWald, PSubRao, PASD, and ASD is plotted for different K. We can see that the proposed persymmetric detectors outperform the ASD. Both the simulated data and real data results demonstrate the effectiveness of the proposed PSubGLRT, PSubWald, PSubRao.

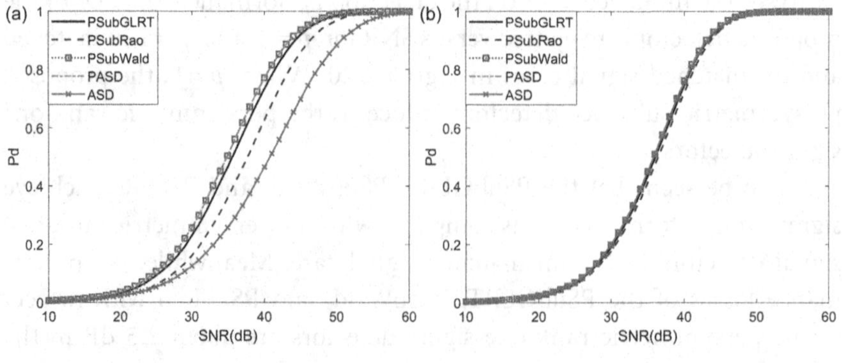

FIGURE 2.21 PD versus SNR for $N=4, p=3$. (a) $K=4$; (b) $K=8$.

2.6 CONCLUSION

In this chapter, we considered the problem of signal detection in a sample-starved environment. To solve the problem, we adopted three kinds of approaches, namely, model-based approach, persymmetric-based approach, and orthogonal-partition-based approach.

For the model-based approach, the MAMF and MRao were derived for the point-like target and distributed target by modelling the noise as an AR process. When the number of pulses is moderate, the MAMF and the MRao can achieve a satisfactory detection performance even in the sample-starved scenario.

For the persymmetric-based approach, the PRao and PWald were designed for the distributed target by exploiting the persymmetric structure of the noise covariance matrix. The PRao and PWald outperform the conventional Rao and Wald detectors since the prior information of the noise is used. The PSubGLRT, PSubRao, and PSubWald were derived for the subspace signals in the PHE. The PSubGLRT, PSubRao, and PSubWald achieve better detection performance than the ASD and are more robust than the persymmetric rank-one signal detectors.

Under the line of orthogonal partition, we proposed the OP-RR-GANMF by projecting the data into several orthogonal subspaces. The OP-RR-GANMF is in the form of the ratio of the signal energy in the signal subspace with clutter being suppressed and the product of the total energy of the test data in the noise subspace and the total energy of the training data in the noise subspace. A main feature is that, unlike the GASD, the OP-RR-GANMF can work well in a low sample support environment even when the number of the sample data is lower than that of the dimension of the signal.

APPENDIX 2.A THE CIRCULARLY SYMMETRIC COMPLEX GAUSSIAN VECTOR

The vector $z = x + jy$ is a complex Gaussian random vector when x and y are jointly Gaussian random vectors. The complex Gaussian random vector z is circularly symmetric if z has the same distribution as $e^{j\phi}z$ for all real ϕ. The circularly symmetric complex Gaussian vector z has the following properties: $E\left[\mathbf{z}\right] = \mathbf{0}, E\left[\mathbf{z}\mathbf{z}^T\right] = \mathbf{0}$ [55].

It has been proved that a complex Gaussian vector is circularly symmetric if and only if its mean and pseudocovariance are zero, with the pseudocovariance defined as $E\left[\left(\mathbf{z} - E\left[\mathbf{z}\right]\right)\left(\mathbf{z} - E\left[\mathbf{z}\right]\right)^T\right] = \mathbf{C}_x - \mathbf{C}_y + j\left(\mathbf{C}_{xy} + \mathbf{C}_{yx}\right)$.

For a circularly symmetric complex Gaussian vector, the following equations hold: $\mathbf{C}_x = \mathbf{C}_y$, $\mathbf{C}_{xy} = -\mathbf{C}_{yx}$. The PDF of the circularly symmetric complex Gaussian vector z is [55, 56]

$$p(\mathbf{z}) = \frac{1}{\pi^N \det(\mathbf{C}_\mathbf{z})} \exp\left(-\mathbf{z}^H \mathbf{C}_\mathbf{z}^{-1} \mathbf{z}\right) \tag{2.169}$$

where $\mathbf{C}_\mathbf{z} = E\left[\left(\mathbf{z} - E[\mathbf{z}]\right)\left(\mathbf{z} - E[\mathbf{z}]\right)^H\right]$.

APPENDIX 2.B THE DERIVATION OF (2.87)

In this appendix, we give the brief mathematical derivations of Equation 2.87. The PDF of $\boldsymbol{X}$, under hypothesis H_1, is given by

$$f_1(\boldsymbol{X}) = \frac{\exp\left\{-\mathrm{tr}\left[(\boldsymbol{X} - \boldsymbol{s}\boldsymbol{a}^H)^H \boldsymbol{R}^{-1} (\boldsymbol{X} - \boldsymbol{s}\boldsymbol{a}^H)\right] / \sigma^2\right\}}{\pi^{NK} \sigma^{2NK} |\boldsymbol{R}|^K} \tag{2.170}$$

Taking the derivative of Equation 2.170 with respect to (w.r.t.) $\boldsymbol{a}$ and equating the result to zero, yields the MLE of $\boldsymbol{a}$ as:

$$\hat{\boldsymbol{a}} = \frac{\boldsymbol{X}^H \boldsymbol{R}^{-1} \boldsymbol{s}}{\boldsymbol{s}^H \boldsymbol{R}^{-1} \boldsymbol{s}} \tag{2.171}$$

Plugging Equation 2.171 into Equation 2.170, after some algebra, results in

$$f_1(\boldsymbol{X}; \hat{\boldsymbol{a}}) = \frac{\exp\left[-\mathrm{tr}\left(\bar{\boldsymbol{X}}^H \boldsymbol{P}_{\bar{s}}^{\perp} \bar{\boldsymbol{X}}\right) / \sigma^2\right]}{\pi^{NK} \sigma^{2NK} |\boldsymbol{R}|^K} \tag{2.172}$$

where $\boldsymbol{P}_{\bar{s}}^{\perp}$ is the orthogonal complement of $\boldsymbol{P}_{\bar{s}}$, i.e., $\boldsymbol{P}_{\bar{s}}^{\perp} = \boldsymbol{I}_N - \boldsymbol{P}_{\bar{s}}$.

Taking the derivative of the logarithm of Equation 2.172 w.r.t. σ^2 and setting the result to be zero, yields the MLE of σ^2, described as

$$\hat{\sigma}_1^2 = \mathrm{tr}\left(\bar{\boldsymbol{X}}^H \boldsymbol{P}_{\bar{s}}^{\perp} \bar{\boldsymbol{X}}\right) / K \tag{2.173}$$

Inserting Equation 2.173 into Equation 2.172, we have the following result

$$f_1(\boldsymbol{X}; \hat{\boldsymbol{a}}, \hat{\sigma}_1^2) = \frac{\left[\mathrm{tr}\left(\bar{\boldsymbol{X}}^H \boldsymbol{P}_{\bar{s}}^{\perp} \bar{\boldsymbol{X}}\right)\right]^{-K}}{\left(e\pi^N |\boldsymbol{R}| / K\right)^K} \tag{2.174}$$

The PDF of $\boldsymbol{X}$, under hypothesis H_0, is given by

$$f_0(\boldsymbol{X}) = \frac{\exp\left[-\mathrm{tr}\left(\boldsymbol{X}^H \boldsymbol{R}^{-1} \boldsymbol{X}\right)/\sigma^2\right]}{\pi^{NK} \sigma^{2NK} \left|\boldsymbol{R}\right|^K} \tag{2.175}$$

In a manner similar to the derivation of Equation 2.174, we can readily obtain the result

$$f_0\left(\boldsymbol{X}; \hat{\sigma}_0^2\right) = \frac{\left[\mathrm{tr}\left(\bar{\boldsymbol{X}}^H \boldsymbol{P}_{\bar{s}} \bar{\boldsymbol{X}}\right)\right]^{-K}}{\left(e\pi^N \left|\boldsymbol{R}\right|/K\right)^K} \tag{2.176}$$

Taking the K th root of the ratio of Equations 2.174 and 2.176, we have the GLRT as:

$$t'_{\mathrm{GLRT}} = \frac{\mathrm{tr}\left(\bar{\boldsymbol{X}}^H \boldsymbol{P}_{\bar{s}} \bar{\boldsymbol{X}}\right)}{\mathrm{tr}\left(\bar{\boldsymbol{X}}^H \boldsymbol{P}_{\bar{s}}^{\perp} \bar{\boldsymbol{X}}\right)} \tag{2.177}$$

which is statistically equivalent to

$$t''_{\mathrm{GLRT}} = \frac{\mathrm{tr}\left(\bar{\boldsymbol{X}}^H \boldsymbol{P}_{\bar{s}} \bar{\boldsymbol{X}}\right)}{\mathrm{tr}\left(\bar{\boldsymbol{X}}^H \bar{\boldsymbol{X}}\right)} \tag{2.178}$$

since $t''_{\mathrm{GLRT}} = [1 + (t'_{\mathrm{GLRT}})^{-1}]^{-1}$ is a monotonically increasing function of t'_{GLRT}. Equation 2.178 is essentially Equation 2.87. This completes the proof.

NOTE

1. It should be pointed out that in some references, whether σ_n^2 equals 1 is considered as the key to distinguish whether the environment is homogenous. However, the fact whether σ_n^2 is known or not is real the key.

REFERENCES

1. W. Liu, J. Liu, C. Hao, Y. Gao, and Y. Wang, "Multichannel Adaptive Signal Detection: Basic Theory and Literature Review," *Science China Information Sciences*, vol. 65, no. 2, pp. 1–41, 2022.
2. O. Besson, "Adaptive Detection Using Randomly Reduced Dimension Generalized Likelihood Ratio Test," *Signal Processing*, vol. 166, pp. 1–4, 2020.

3. Z. Wang, Z. Zhao, C. Ren, Z. Nie, and W. Yang, "Adaptive Detection of Point-Like Targets Based on a Reduced-Dimensional Data Model," *Signal Processing*, vol. 158, pp. 36–47, 2019.
4. J. Lin, C. Jiang, J. Jiang, and J. Kang, "Conjugate Gradient Persymmetric Adaptive Matched Filter," *Digital Signal Processing*, vol. 123, pp. 1–10, 2022.
5. H. Li, W. Song, W. Liu, and R. Wu, "Moving Target Detection with Limited Secondary Data Based on the Subspace Orthogonal Projection," *IET Radar Sonar Navigation*, vol. 12, pp. 679–684, 2018.
6. Y. Gao, H. Li, and B. Himed, "Adaptive Subspace Tests for Multichannel Signal Detection in Auto-Regressive Noise," *IEEE Transactions on Signal Processing*, vol. 66, no. 21, pp. 5577–5587, 2018.
7. L. Yan, C. Hao, D. Orlando, A. Farina, and C. Hou, "Parametric Space-Time Detection and Range Estimation of Point-Like Targets in Partially Homogeneous Environment," *IEEE Transactions on Aerospace and Electronic Systems*, vol. 56, no. 2, pp. 1228–1242, 2020.
8. J. Xue, S. Xu, J. Liu, M. Pan, and J. Fang, "Bayesian Detection for Radar Targets in Compound-Gaussian Sea Clutter," *IEEE Geoscience and Remote Sensing Letters*, vol. 19, pp. 1–5, 2022.
9. Z. Wang, J. Liu, Y. Li, H. Chen, and M. Peng, "Adaptive Subspace Signal Detection in Structured Interference Plus Compound Gaussian Sea Clutter," *Remote Sensing*, vol. 14, pp. 1–13, 2022.
10. X. Du, A. Aubry, A. De Maio, and G. Cui, "Toeplitz Structured Covariance Matrix Estimation for Radar Applications," *IEEE Signal Processing Letters*, vol. 27, pp. 595–599, 2020.
11. L. Xie, Z. He, J. Tong, T. Liu, J. Li, and J. Xi, "Regularized Covariance Estimation for Polarization Radar Detection in Compound Gaussian Sea Clutter," *IEEE Transactions on Geoscience and Remote Sensing*, vol. 60, pp. 1–16, 2022.
12. G. Alfano, A. De Maio, and A. Farina, "Model-Based Adaptive Detection of Range-Spread Targets," *IET Radar Sonar Navigation*, vol. 151, no. 1, pp. 2–10, 2004.
13. S. M. Kay, *Fundamentals of Statistical Signal Processing, Volume II: Detection Theory*. Prentice-Hall, Englewood Cliffs, NJ, 1998.
14. A. Sheikhi, M. M. Nayebi, and M. R. Aref, "Adaptive Detection Algorithm for Radar Signals in Autoregressive Interference," *IET Radar Sonar Navigation*, vol. 145, no. 5, pp. 309–314, 1998.
15. X. Shuai, L. Kong, and J. Yang, "AR-Model-Based Adaptive Detection of Range-Spread Targets in Compound Gaussian Clutter," *Signal Processing*, vol. 91, pp. 750–758, 2011.
16. J. Li, B. Halder, P. Stoica, and M. Viberg, "Computationally Efficient Angle Estimation for Signals with Known Waveforms," *IEEE Transactions on Signal Processing*, vol. 43, no. 9, pp. 2154–2163, 1995.
17. S. M. Kay, *Modern Spectral Estimation: Theory & Application*. Prentice-Hall, Englewood Cliffs, NJ, 1988.

18. E. Conte and A. De Maio, "Distributed Target Detection in Compound-Gaussian Noise with Rao and Wald Tests," *IEEE Transactions on Aerospace and Electronic Systems*, vol. 39, no. 2, pp. 568–582, 2003.
19. E. Conte, A. D. Maio, and G. Ricci, "GLRT-Based Adaptive Detection Algorithms for Range-Spread Targets," *IEEE Transactions on Signal Processing*, vol. 49, no. 7, pp. 1336–1348, 2001.
20. Y.-L. Gau and I. S. Reed, "An Improved Reduced-Rank CFAR Space-Time Adaptive Radar Detection Algorithm," *IEEE Transactions on Signal Processing*, vol. 46, no. 8, pp. 2139–2146, 1998.
21. W. Liu, W. Xie, and Y.-L. Wang, "AMF And ACE Detectors Based on Diagonal Loading," *Systems Engineering and Electronics*, vol. 35, no. 3, pp. 463–468, 2013.
22. S. Kraut and L. L. Scharf, "The CFAR Adaptive Subspace Detector Is A Scale-Invariant GLRT," *IEEE Transactions on Signal Processing*, vol. 47, no. 9, pp. 2538–2541, 1999.
23. I. S. Reed, Y. L. Gau, and T. K. Truong, "CFAR Detection and Estimation for STAP Radar," *IEEE Transactions on Aerospace and Electronic Systems*, vol. 34, no. 3, pp. 722–735, 1998.
24. U. Nickel, "Design of Generalised 2D Adaptive Sidelobe Blanking Detectors Using The Detection Margin," *Signal Processing*, vol. 90, no. 5, pp. 1357–1372, 2010.
25. Y.-L. Wang, W. Liu, W. Xie, and Y. Zhao, "Reduced-Rank Space-Time Adaptive Detection For Airborne Radar," *Science China: Information Sciences*, vol. 57, no. 8, pp. 1–11, 2014.
26. W. Liu, W. Xie, and Y. Wang, "Some Complex Statistical Distributions in Complex-Valued Signal Detection Theory," *Acta Electronica Sinica*, vol. 41, no. 6, pp. 1238–1241, 2013.
27. M. Casillo, A. De Maio, S. Iommelli, and L. Landi, "A Persymmetric GLRT for Adaptive Detection in Partially-Homogeneous Environment," *IEEE Signal Processing Letters*, vol. 14, no. 12, pp. 1016–1019, 2007.
28. C. Hao, D. Orlando, X. Ma, and C. Hou, "Persymmetric Rao and Wald Tests for Partially Homogeneous Environment," *IEEE Signal Processing Letters*, vol. 19, no. 9, pp. 587–590, 2012.
29. E. Conte and A. De Maio, "Exploiting Persymmetry for CFAR Detection in Compound-Gaussian Noise," *IEEE Transactions on Aerospace and Electronic Systems*, vol. 39, no. 2, pp. 719–724, 2003.
30. C. Hao, D. Orlando, G. Foglia, X. Ma, S. Yan, and C. Hou, "Persymmetric Adaptive Detection of Distributed Targets in Partially-Homogeneous Environment," *Digital Signal Processing*, vol. 24, pp. 42–51, 2014.
31. A. De Maio and D. Orlando, "An Invariant Approach to Adaptive Radar Detection Under Covariance Persymmetry," *IEEE Transactions on Signal Processing*, vol. 63, no. 5, pp. 1297–1309, 2015.
32. J. Liu, H. Li, and B. Himed, "Persymmetric Adaptive Target Detection with Distributed MIMO Radar," *IEEE Transactions on Aerospace and Electronic Systems*, vol. 51, no. 1, pp. 372–382, 2015.

33. J. Xue, H. Li, M. Pan, and J. Liu, "Adaptive Persymmetric Detection for Radar Targets in Correlated CG-LN Sea Clutter," *IEEE Transactions on Geoscience and Remote Sensing*, vol. 61, pp. 1–12, 2023.
34. K. Gerlach and M. J. Steiner, "Adaptive Detection of Range Distributed Targets," *IEEE Transactions on Signal Processing*, vol. 47, no. 7, pp. 1844–1851, 1999.
35. N. B. Pulsone and R. S. Raghavan, "Analysis of An Adaptive CFAR Detector in Non-Gaussian Interference," *IEEE Transactions on Aerospace and Electronic Systems*, vol. 35, no. 3, pp. 903–916, 1999.
36. G. A. Fabrizio, A. Farina, and M. D. Turley, "Spatial Adaptive Subspace Detection in OTH Radar," *IEEE Transactions on Aerospace and Electronic Systems*, vol. 39, no. 4, pp. 1407–1428, 2003.
37. C. Hao, X. Ma, X. Shang, and L. Cai, "Adaptive Detection of Distributed Targets in Partially Homogeneous Environment with Rao and Wald Tests," *Signal Processing*, vol. 92, no. 4, pp. 926–930, 2012.
38. N. B. Pulsone and C. M. Rader, "Adaptive Beamformer Orthogonal Rejection Test," *IEEE Transactions on Signal Processing*, vol. 49, no. 3, pp. 521–529, Mar. 2001.
39. R. Bakker and B. Currie, The McMaster IPIX Radar Sea Noise Database, http://soma.ece.mcmaster.ca/ipix/.
40. D. Mata-Moya, N. del-Rey-Maestre, M. P. Jarabo-Amores, J. Martin-de-Nicolas, and J. L. Barcena-Humanes, "An Adaptive Threshold Technique for The LR Detector In K-Noise: Validation Using IPIX Radar," *IEEE Instrumentation and Measurement Technology Conference*, Pisa, Italy, pp. 794–799, May 2015.
41. E. Conte, A. De Maio, and A. Farina, "Statistical Tests for Higher Order Analysis of Radar Noise: Their Application To L-Band Measured Data," *IEEE Transactions on Aerospace and Electronic Systems*, vol. 41, no. 1, pp. 205–218, 2005.
42. O. Besson, L. L. Scharf, and F. Vincent. "Matched Direction Detectors and Estimators for Array Processing with Subspace Steering Vector Uncertainties," *IEEE Transactions on Signal Processing*, vol. 53, no. 12, pp. 4453–4463, 2005.
43. Y. L. Dong, M. Liu, K. Li, Z. K. Tang, and W. J. Liu, "Adaptive Direction Detection in Deterministic Interference and Partially Homogeneous Noise," *IEEE Signal Processing Letters*, vol. 24, no. 5, pp. 599–603, May 2017.
44. F. Gini and A. Farina, "Vector Subspace Detection in Compound-Gaussian Noise Part I: Survey and New Results," *IEEE Transactions on Aerospace and Electronic Systems*, vol. 38, no. 4, pp. 1295–1311, 2002.
45. J. Liu and J. Li, "False Alarm Rate of the GLRT for Subspace Signals in Subspace Interference Plus Gaussian Noise," *IEEE Transactions on Signal Processing*, vol. 67, no. 11, pp. 3058–3069, 2019.
46. A. De Maio, G. Alfano, and E. Conte, "Polarization Diversity Detection in Compound-Gaussian Noise," *IEEE Transactions on Aerospace and Electronic Systems*, vol. 40, no. 1, pp. 114–131, 2004.

47. J. Liu and J. Li, "Robust Detection in MIMO Radar with Steering Vector Mismatches," *IEEE Transactions on Signal Processing*, vol. 67, no. 20, pp. 5270–5280, 2019.
48. W. J. Liu, W. C. Xie, J. Liu, and Y. L. Wang, "Adaptive Double Subspace Signal Detection in Gaussian Background-Part II: Partially Homogeneous Environments," *IEEE Transactions on Signal Processing*, vol. 62, no. 9, pp. 2358–2369, 2014.
49. J. Liu and J. Li, "Mismatched Signal Rejection Performance of The Persymmetric GLRT Detector," *IEEE Transactions on Signal Processing*, vol. 67, no. 6, pp. 1610–1619, 2019.
50. J. Liu, S. Y. Sun, and W. J. Liu, "One-Step Persymmetric GLRT for Subspace Signals," *IEEE Transactions on Signal Processing*, vol. 67, no. 14, pp. 3639–3648, 2019.
51. A. Hjørungnes, *Complex-Valued Matrix Derivatives: With Applications in Signal Processing and Communications*, Cambridge University Press, Cambridge, 2011.
52. E. J. Kelly and K. Forsythe, *Adaptive Detection and Parameter Estimation for Multidimensional Signal Models*, Lexington: Lincoln Laboratory, Tech. Rep. 848, 1989.
53. J. Liu, W. J. Liu, Y. C. Gao, S. H. Zhou, and X. G. Xia, "Persymmetric Adaptive Detection of Subspace Signals: Algorithms and Performance Analysis," *IEEE Transactions on Signal Processing*, vol. 66, no. 23, pp. 6124–6136, 2018.
54. Y. C. Gao, G. S. Liao, S. Q. Zhu, X. P. Zhang, and D. Yang, "Persymmetric Adaptive Detectors in Homogeneous and Partially Homogeneous Environments," *IEEE Transactions on Signal Processing*, vol. 62, no. 2, pp. 331–342, 2014.
55. http://en.wikipedia.org/wiki/Complex_normal_distribution.
56. F. Bandiera, D.Orlando, and G. Ricci, *Advanced Radar Detection Schemes Under Mismatched Signal Models*. Morgan & Claypool Publishers, San Rafael, CA, USA, 2009.

CHAPTER 3

Adaptive Selective Detectors for Mismatched Signals in the Presence of Signal Mismatch

As discussed in Chapter 1, non-ideal factors like sidelobe interference, uncalibrated arrays, and beam pointing errors will cause the deviation of the nominal steering vector from the actual steering vector. In signal mismatch cases, the selective detectors have lower probabilities of detection for the mismatched signals. The conventional detectors derived under the assumption that the actual target steering vector is exactly the nominal steering vector exhibit different selectivities to mismatched signals. The ACE, originally proposed in the PHE in [1], is more selective than the AMF and Kelly's GLRT.

There are two commonly used methods to enhance the selectivity of the adaptive detectors. One method is to modify the binary hypothesis testing by introducing an unwanted signal under the null hypothesis. In [2], a signal that is orthogonal to the nominal steering vector in the quasi-whitened space is added under the null hypothesis. An adaptive detector is derived

 DOI: 10.1201/9781003477907-3

by resorting to the GLRT decision rule in Gaussian noise. The resulting adaptive beamformer orthogonal rejection test (ABORT) achieves stronger mismatch rejection capability and similar matched detection performance as Kelly's GLRT and AMF. By introducing fictitious signals that are orthogonal to the nominal steering vector in the truly whitened observation space under the null hypothesis, an abort-like detector is derived in [3]. The abort-like detector achieves a better mismatched signal discrimination performance than the ABORT at the price of a certain matched signal detection performance. In [4], the ABORT is extended to the subspace target case wherein the target is described by the linear subspace model, and an unwanted signal that lies in a subspace orthogonal to the target subspace is introduced. In [5], the ABORT is extended to the case where a fictitious signal under the null hypothesis is modelled probabilistically. Meanwhile, a Bayesian framework is designed to cope with the limited training samples problem. The other method that can enhance the selectivity of the mismatched signals is to add a random noise-like interferer to the CUT. In [6], the CUT is assumed to contain the noise-like interferer, noise, clutter, and possible target signal and a Rao test with enhanced selectivity with respect to the abort-like detector is derived.

In this chapter, the design of the selective detectors is discussed. In Section 3.1, the design of two selective detectors based on the one-step GLRT and two-step GLRT in PHE is given. In Section 3.2, two-step GLRT and MAP GLRT are resorted to design adaptive detectors with enhanced selectivity in CG noise.

3.1 ADAPTIVE SELECTIVE DETECTORS IN PARTIALLY HOMOGENEOUS ENVIRONMENT

3.1.1 Problem Formulation

Adaptive detection of a signal in PHE is discussed. To improve the selectivity, a fictitious unwanted signal is added under the null hypothesis. The detection problem is then transformed into deciding whether received data contain a useful target signal or an unwanted signal. The binary hypotheses testing becomes

$$\begin{cases} H_0 : \boldsymbol{y}_0 = \beta \boldsymbol{u} + \boldsymbol{n}_0, & \boldsymbol{y}_l = \boldsymbol{n}_l, l = 1,\ldots,K \\ H_1 : \boldsymbol{y}_0 = \alpha \boldsymbol{v} + \boldsymbol{n}_0, & \boldsymbol{y}_l = \boldsymbol{n}_l, l = 1,\ldots,K \end{cases} \tag{3.1}$$

where y_0 denotes the data under test, namely, the primary data, $y_l \in \mathbb{C}^{N\times 1}$, $l=1,\ldots,K$ denote the training data, $v \in \mathbb{C}^{N\times 1}$ is the nominal steering vector, α is the unknown target amplitude, βu denotes the fictitious unwanted signal, which is orthogonal to the nominal steering vector in quasi-whitened observation space, $\left\langle S^{-\frac{1}{2}}u \right\rangle = \left\langle S^{-\frac{1}{2}}v \right\rangle^{\perp}$, $S = \sum_{l=1}^{K} y_l y_l^H$, $\left\langle S^{-\frac{1}{2}}u \right\rangle$ denotes the space spanned by the columns of $S^{-\frac{1}{2}}u$, $\left\langle S^{-\frac{1}{2}}v \right\rangle^{\perp}$ denotes the orthogonal complement of $\left\langle S^{-\frac{1}{2}}v \right\rangle$, β is the unknown coordinate, the noise $n_0 \in \mathbb{C}^{N\times 1}$ and $n_l \in \mathbb{C}^{N\times 1}, l=1,\ldots,K$ are zero-mean complex Gaussian vectors, $E\left[n_0 n_0^H\right] = M$, $E\left[n_l n_l^H\right] = \gamma M$, $\gamma > 0$.

3.1.2 One-Step Selective Detector

The joint probability density functions of the data under test and training data under H_0 and H_1 are

$$f(Y \mid H_0) = \frac{\gamma^{-NK}}{\pi^{N(K+1)} \det^{(K+1)}(M)} \exp\left\{-\mathrm{tr}\left[M^{-1}\left((y_0 - \beta u)(y_0 - \beta u)^H + \frac{1}{\gamma}S\right)\right]\right\} \tag{3.2}$$

$$f(Y \mid H_1) = \frac{\gamma^{-NK}}{\pi^{N(K+1)} \det^{(K+1)}(M)} \exp\left\{-\mathrm{tr}\left[M^{-1}\left((y_0 - \alpha v)(y_0 - \alpha v)^H + \frac{1}{\gamma}S\right)\right]\right\} \tag{3.3}$$

where $Y = \left[y_0 \ldots, y_K\right]$ denote data under test and the training data.

For the detection problem in Equation 3.1, the generalized likelihood ratio test is expressed as

$$\frac{\max\limits_{\alpha} \max\limits_{\gamma} \max\limits_{M} f(Y \mid H_1)}{\max\limits_{\beta} \max\limits_{\gamma} \max\limits_{M} f(Y \mid H_0)} \underset{H_0}{\overset{H_1}{\gtrless}} \eta \tag{3.4}$$

Maximizing the PDF in Equation 3.2 with respect to $\boldsymbol{M}$, we get the MLE of $\boldsymbol{M}$ under H_0

$$\hat{\boldsymbol{M}}_0 = \left[(\boldsymbol{y}_0 - \beta \boldsymbol{u})(\boldsymbol{y}_0 - \beta \boldsymbol{u})^H + \frac{1}{\gamma}\boldsymbol{S} \right] \Big/ (K+1) \tag{3.5}$$

By plugging Equation 3.5 into Equation 3.2, the MLE of β can be calculated as follows:

$$\hat{\beta} = \arg\min_{\beta} \det \left[(\boldsymbol{y}_0 - \beta \boldsymbol{u})(\boldsymbol{y}_0 - \beta \boldsymbol{u})^H + \frac{1}{\gamma}\boldsymbol{S} \right] = \frac{\boldsymbol{u}^H \boldsymbol{S}^{-1} \boldsymbol{y}_0}{\boldsymbol{u}^H \boldsymbol{S}^{-1} \boldsymbol{u}} \tag{3.6}$$

We substitute Equations 3.5 and 3.6 into Equation 3.2 and obtain the PDF $f\left(\boldsymbol{Y} | \hat{M}_0, \hat{\beta}, H_0\right)$

$$f\left(\boldsymbol{Y} | \hat{\mathrm{M}}_0, \hat{\beta}, H_0\right) \propto \frac{\gamma^{-NK}}{\det^{(K+1)}\left[\left(\boldsymbol{y}_0 - \hat{\beta}\boldsymbol{u}\right)\left(\boldsymbol{y}_0 - \hat{\beta}\boldsymbol{u}\right)^H + \frac{1}{\gamma}\boldsymbol{S} \right]} \tag{3.7}$$

From Equation 3.7, the MLE of γ can be computed by minimizing $\gamma^{NK/(K+1)} \det\left[\left(\boldsymbol{y}_0 - \hat{\beta}\boldsymbol{u}\right)\left(\boldsymbol{y}_0 - \hat{\beta}\boldsymbol{u}\right)^H + \frac{1}{\gamma}\boldsymbol{S} \right]$ with respect to γ. Using the determinant identity $\det\left(\frac{1}{\gamma}\boldsymbol{I}_N + \boldsymbol{AB}\right) = \gamma^{L-N} \det\left(\frac{1}{\gamma}\boldsymbol{I}_L + \boldsymbol{BA}\right)$ for $A \in \mathbb{C}^{N \times L}$ and $B \in \mathbb{C}^{L \times N}$, $\gamma^{NK/(K+1)} \det\left[\left(\boldsymbol{y}_0 - \hat{\beta}\boldsymbol{u}\right)\left(\boldsymbol{y}_0 - \hat{\beta}\boldsymbol{u}\right)^H + \frac{1}{\gamma}\boldsymbol{S} \right]$ can be further simplified as

$$\begin{aligned} &\gamma^{NK/(K+1)} \det\left[\left(\boldsymbol{y}_0 - \hat{\beta}\boldsymbol{u}\right)\left(\boldsymbol{y}_0 - \hat{\beta}\boldsymbol{u}\right)^H + \frac{1}{\gamma}\boldsymbol{S} \right] \\ &= \gamma^{NK/(K+1)} \det(\boldsymbol{S}) \det\left[\boldsymbol{S}^{-1} \left(\boldsymbol{y}_0 - \hat{\beta}\boldsymbol{u}\right)\left(\boldsymbol{y}_0 - \hat{\beta}\boldsymbol{u}\right)^H + \frac{1}{\gamma}\boldsymbol{I}_N \right] \\ &= \det(\boldsymbol{S}) \gamma^{[1-N/(K+1)]} \left[\frac{1}{\gamma} + \left(\boldsymbol{y}_0 - \hat{\beta}\boldsymbol{u}\right)^H \boldsymbol{S}^{-1} \left(\boldsymbol{y}_0 - \hat{\beta}\boldsymbol{u}\right) \right] \end{aligned} \tag{3.8}$$

Taking the derivative of Equation 3.8 with respect to γ and setting it to zero, we obtain the MLE of γ under H_0

$$\hat{\gamma}_0 = \frac{N}{(K+1-N)\left[\left(\boldsymbol{y}_0 - \hat{\beta}\boldsymbol{u}\right)^H \boldsymbol{S}^{-1}\left(\boldsymbol{y}_0 - \hat{\beta}\boldsymbol{u}\right)\right]} \tag{3.9}$$

In a similar way, we can get the MLEs of the unknown parameters under H_1

$$\hat{\gamma}_1 = \frac{N}{(K+1-N)\left[\left(\boldsymbol{y}_0 - \hat{\alpha}\boldsymbol{v}\right)^H \boldsymbol{S}^{-1}\left(\boldsymbol{y}_0 - \hat{\alpha}\boldsymbol{v}\right)\right]} \tag{3.10}$$

$$\hat{\alpha} = \frac{\boldsymbol{v}^H \boldsymbol{S}^{-1}\boldsymbol{y}_0}{\boldsymbol{v}^H \boldsymbol{S}^{-1}\boldsymbol{v}} \tag{3.11}$$

The condition that $\boldsymbol{u} \in \mathbb{C}^{N\times 1}$ is orthogonal to the nominal steering vector in quasi-whitened observation space implies

$$\boldsymbol{I} - \frac{\boldsymbol{S}^{-\frac{1}{2}}\boldsymbol{u}\boldsymbol{u}^H\boldsymbol{S}^{-\frac{1}{2}}}{\boldsymbol{u}^H\boldsymbol{S}^{-1}\boldsymbol{u}} = \frac{\boldsymbol{S}^{-\frac{1}{2}}\boldsymbol{v}\boldsymbol{v}^H\boldsymbol{S}^{-\frac{1}{2}}}{\boldsymbol{v}^H\boldsymbol{S}^{-1}\boldsymbol{v}} \tag{3.12}$$

Then, the expression $\left(\boldsymbol{y}_0 - \hat{\beta}\boldsymbol{u}\right)^H \boldsymbol{S}^{-1}\left(\boldsymbol{y}_0 - \hat{\beta}\boldsymbol{u}\right)$ can be rewritten as

$$\begin{aligned} &\left(\boldsymbol{y}_0 - \hat{\beta}\boldsymbol{u}\right)^H \boldsymbol{S}^{-1}\left(\boldsymbol{y}_0 - \hat{\beta}\boldsymbol{u}\right) \\ &= \boldsymbol{y}_0^H \boldsymbol{S}^{-\frac{1}{2}}\left(\boldsymbol{I} - \frac{\boldsymbol{S}^{-\frac{1}{2}}\boldsymbol{u}\boldsymbol{u}^H\boldsymbol{S}^{-\frac{1}{2}}}{\boldsymbol{u}^H\boldsymbol{S}^{-1}\boldsymbol{u}}\right)\boldsymbol{S}^{-\frac{1}{2}}\boldsymbol{y}_0 \\ &= \frac{\boldsymbol{y}_0^H\boldsymbol{S}^{-1}\boldsymbol{v}\boldsymbol{v}^H\boldsymbol{S}^{-1}\boldsymbol{y}_0}{\boldsymbol{v}^H\boldsymbol{S}^{-1}\boldsymbol{v}} \end{aligned} \tag{3.13}$$

Finally, plugging Equations 3.5–3.13 into Equation 3.4, we obtain the mismatch selective one-step GLRT (SE-1SGLRT)

$$\frac{\boldsymbol{y}_0^H\boldsymbol{S}^{-\frac{1}{2}}\boldsymbol{P}_{\boldsymbol{v}_s}\boldsymbol{S}^{-\frac{1}{2}}\boldsymbol{y}_0}{\boldsymbol{y}_0^H\boldsymbol{S}^{-\frac{1}{2}}\left(\boldsymbol{I}_N - \boldsymbol{P}_{\boldsymbol{v}_s}\right)\boldsymbol{S}^{-\frac{1}{2}}\boldsymbol{y}_0} \underset{H_0}{\overset{H_1}{\gtrless}} \eta_{SE-1SGLRT} \tag{3.14}$$

where $\boldsymbol{P}_{\boldsymbol{v}_s} = \frac{\boldsymbol{S}^{-\frac{1}{2}}\boldsymbol{v}\boldsymbol{v}^H\boldsymbol{S}^{-\frac{1}{2}}}{\boldsymbol{v}^H\boldsymbol{S}^{-1}\boldsymbol{v}}$, $\eta_{SE-1SGLRT}$ denotes the transformation of the threshold η.

After some calculation, the mismatch selective one-step GLRT (3.14) can be reformulated as

$$\frac{1}{1/t_{ACE}-1} \underset{H_0}{\overset{H_1}{\gtrless}} \eta_{SE-1SGLRT} \tag{3.15}$$

where $t_{ACE} = \frac{\left|\boldsymbol{y}_0^H\boldsymbol{S}^{-1}\boldsymbol{v}\right|^2}{\left(\boldsymbol{v}^H\boldsymbol{S}^{-1}\boldsymbol{v}\right)\left(\boldsymbol{y}_0^H\boldsymbol{S}^{-1}\boldsymbol{y}_0\right)}$. From Equation 3.15, it can be seen that the test statistic of the mismatch selective one-step GLRT is proportional to that of the ACE derived in [1].

3.1.3 Two-Step Selective Detector

The two-step design criterion is resorted to solve the detection problem in Equation 3.1. The noise covariance matrix $\boldsymbol{M}$ is first assumed known and the one-step GLRT is derived. Then, the estimate of $\boldsymbol{M}$ is substituted into the test statistic to obtain a completely adaptive detector.

The GLRT with known $\boldsymbol{M}$ is:

$$\frac{\max_{\alpha} f\left(\boldsymbol{y}_0 \mid H_1\right)}{\max_{\beta} f\left(\boldsymbol{y}_0 \mid H_0\right)} \underset{H_0}{\overset{H_1}{\gtrless}} \eta_{SE-2SGLRT} \tag{3.16}$$

where $\eta_{SE-2SGLRT}$ is the threshold, $f\left(\boldsymbol{y}_0 \mid H_i\right)$ is the PDF of the primary data, i.e.,

$$f\left(\boldsymbol{y}_0 \mid H_0\right) \propto \frac{1}{\det\left(\boldsymbol{M}\right)}\exp\left\{-\mathrm{tr}\left[\boldsymbol{M}^{-1}\left(\boldsymbol{y}_0-\beta\boldsymbol{u}\right)\left(\boldsymbol{y}_0-\beta\boldsymbol{u}\right)^H\right]\right\} \tag{3.17}$$

$$f(\boldsymbol{y}_0 | H_1) \propto \frac{1}{\det(\boldsymbol{M})} \exp\left\{-\mathrm{tr}\left[\boldsymbol{M}^{-1}(\boldsymbol{y}_0 - \alpha\boldsymbol{v})(\boldsymbol{y}_0 - \alpha\boldsymbol{v})^H\right]\right\} \tag{3.18}$$

We can get the MLEs of β and α easily

$$\hat{\beta} = \frac{\boldsymbol{u}^H \boldsymbol{M}^{-1} \boldsymbol{y}_0}{\boldsymbol{u}^H \boldsymbol{M}^{-1} \boldsymbol{u}} \tag{3.19}$$

$$\hat{\alpha} = \frac{\boldsymbol{v}^H \boldsymbol{M}^{-1} \boldsymbol{y}_0}{\boldsymbol{v}^H \boldsymbol{M}^{-1} \boldsymbol{v}} \tag{3.20}$$

Plugging Equations 3.17–3.20 into Equation 3.16 yields the test statistic with known $\boldsymbol{M}$

$$\exp\left(\frac{\boldsymbol{y}_0^H \boldsymbol{M}^{-1} \boldsymbol{v}\boldsymbol{v}^H \boldsymbol{M}^{-1} \boldsymbol{y}_0}{\boldsymbol{v}^H \boldsymbol{M}^{-1} \boldsymbol{v}} - \frac{\boldsymbol{y}_0^H \boldsymbol{M}^{-1} \boldsymbol{u}\boldsymbol{u}^H \boldsymbol{M}^{-1} \boldsymbol{y}_0}{\boldsymbol{u}^H \boldsymbol{M}^{-1} \boldsymbol{u}}\right) \underset{H_0}{\overset{H_1}{\gtrless}} \eta_{SE-2SGLRT} \tag{3.21}$$

In the second step, we estimate M by using the training data

$$\gamma\hat{\boldsymbol{M}} = \frac{1}{K}\sum_{l=1}^{K} \boldsymbol{y}_l \boldsymbol{y}_l^H = \frac{1}{K}\boldsymbol{S} \tag{3.22}$$

Considering the fact that the fictitious signal is orthogonal to the nominal steering vector in quasi-whitened observation space, we substitute Equations 3.12 and 3.22 into Equation 3.21 and obtain the final mismatch selective two-step GLRT (SE-2SGLRT)

$$\exp\left(2\frac{\boldsymbol{y}_0^H \boldsymbol{S}^{-1} \boldsymbol{v}\boldsymbol{v}^H \boldsymbol{S}^{-1} \boldsymbol{y}_0}{\boldsymbol{v}^H \boldsymbol{S}^{-1} \boldsymbol{v}} - \boldsymbol{y}_0^H \boldsymbol{S}^{-1} \boldsymbol{y}_0\right) \underset{H_0}{\overset{H_1}{\gtrless}} \eta_{SE-2SGLRT} \tag{3.23}$$

After some algebra, the mismatch selective two-step GLRT can also be written as

$$\exp\left[y_0^H S^{-\frac{1}{2}} P_{v_s} S^{-\frac{1}{2}} y_0 - y_0^H S^{-\frac{1}{2}} \left(I_N - P_{v_s} \right) S^{-\frac{1}{2}} y_0 \right] \underset{H_0}{\overset{H_1}{\gtrless}} \eta_{SE-2SGLRT} \tag{3.24}$$

3.1.4 Numerical Examples

Monte Carlo simulations are conducted to evaluate the performance of the SE-1SGLRT and the SE-2SGLRT. The PD and the threshold are determined by resorting to $10/P_{fa}$ and $100/P_{fa}$ independent trials, respectively. We set $N = 10$, $\gamma = 2$, $\rho = 0.9$, $P_{fa} = 10^{-3}$. The noise covariance matrix M is $M(i,j) = \rho^{|i-j|}$, $1 \le i, j \le N$. The SNR is defined as $\text{SNR} = |\alpha|^2 v^H M^{-1} v$, where $v = [1, \ldots, 1]^T / \sqrt{N}$. For comparison, the detection performance of the ACE and conventional two-step GLRT (2SGLRT) [7] is also given. The test statistics of the ACE and 2S-GLRT are:

$$t_{ACE} = \frac{\left| y_0^H S^{-1} v \right|^2}{\left(v^H S^{-1} v \right)\left(y_0^H S^{-1} y_0 \right)} \tag{3.25}$$

$$t_{2SGLRT} = \frac{\left| v^H S^{-1} y_0 \right|^2}{\left(v^H S^{-1} v \right)} \tag{3.26}$$

The mismatched detection performance of the SE-1SGLRT and the SE-2SGLRT is assessed first. The signal mismatch is usually measured by the mismatch angle between the actual steering vector v_m and the nominal steering vector v in the whitened observation space, which is defined as [2]: $\cos^2 \phi = \frac{\left| v^H M^{-1} v_m \right|^2}{\left(v^H M^{-1} v \right)\left(v_m^H M^{-1} v_m \right)}$. $\cos^2 \phi = 1$ denotes that the actual steering vector is aligned with the nominal steering vector, while $\cos^2 \phi = 0$ denotes that the actual steering vector is orthogonal to the nominal steering vector. In signal mismatch cases, the SNR becomes $\text{SNR} = |\alpha|^2 v_m^H M^{-1} v_m$.

The contours of const P_d for the SE-1SGLRT, the SE-2SGLRT, the ACE, and the 2S-GLRT are plotted in Figure 3.1. These contour plots, which are first introduced in [2], are also called the mesa plots. Figure 3.1 shows that the SE-1SGLRT and ACE are the most selective. The 2S-GLRT is the

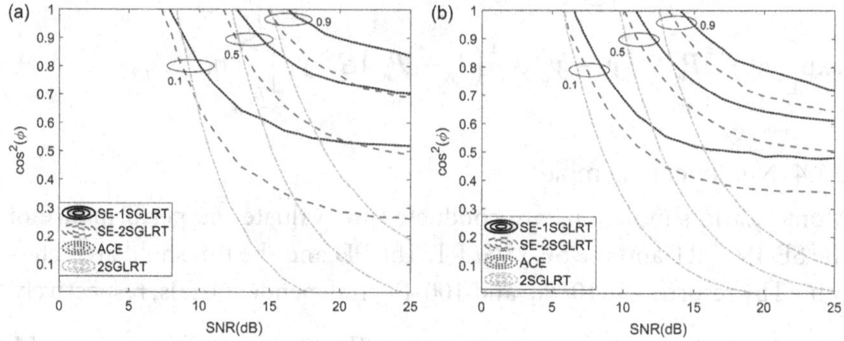

FIGURE 3.1 Contours of const P_d. (a) $K = 20$; (b) $K = 40$.

least sensitive to mismatched signals. The selectivity of the SE-2SGLRT to the mismatched signals is between the SE-1SGLRT and the 2S-GLRT. Meanwhile, the SE-1SGLRT and the ACE coincide, which is consistent with the theoretical analysis.

Figure 3.2 shows the probabilities of detection of SE-1SGLRT, the SE-2SGLRT, the ACE, and the 2S-GLRT against K. From Figure 3.2(a), we can see that the matched detection performance of the SE-2SGLRT is the best. From Figure 3.2(b), we can see that the SE-2SGLRT outperforms the 2S-GLRT when the SNR is low and suffers detection performance degradation when the SNR increases. The matched detection performance loss is about 0.5 dB for $P_d = 0.9$. The detection performance loss of the SE-1SGLRT with respect to the 2S-GLRT is about 1.5 dB for $P_d = 0.9$. Thus, the SE-2SGLRT and SE-1SGLRT have better rejection capabilities

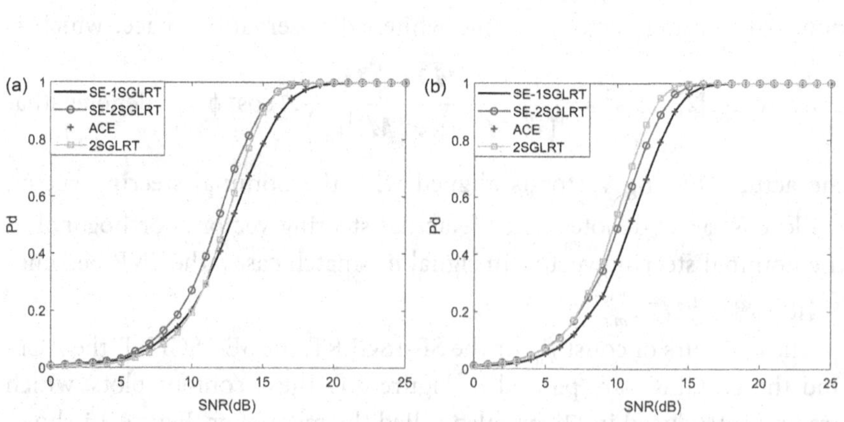

FIGURE 3.2 Probability of detection versus SNR(a) $K = 20$;(b) $K = 40$.

for the mismatched signals but the price is some slight matched detection performance loss.

3.2 ADAPTIVE SELECTIVE DETECTORS IN THE COMPOUND GAUSSIAN ENVIRONMENT

When observed at the low grazing angles or in the high-resolution radar system, the noise may not be Gaussian. The CG model is one of the most widely used non-Gaussian noise models. In the CG model, the noise is expressed as the product of the speckle component and texture component. The texture, which is related to the noise power, is a nonnegative real random variable and the decorrelated length of the texture is usually a few hundred milliseconds [8]. The speckle is a zero-mean complex Gaussian process with a decorrelated length of a few milliseconds. When the texture is a constant, the CG model degrades into the SIRV model. Obviously, the partially homogeneous model wherein the maximum spatial distance of any two range cells to be detected and the maximum spatial distance of any two range cells of the training data are smaller than the correlation distance of the texture component, is a special case of the CG model. In this section, the design of selective detectors in the CG noise is discussed.

3.2.1 Problem Formulation

The received data are assumed to be collected from N channels. We want to determine if the target signal is included in the received data. To enhance the selectivity, we introduce a fictitious unwanted signal $\boldsymbol{F}\boldsymbol{b}_l$ under hypothesis H_0. The detection problem can be expressed as the binary hypothesis testing as follows:

$$\begin{cases} H_0: \begin{cases} \boldsymbol{r}_l = \boldsymbol{F}\boldsymbol{b}_l + \boldsymbol{c}_l, & l = 1,\dots,L \\ \boldsymbol{r}_l = \boldsymbol{c}_l, & l = L+1,\dots,L+K \end{cases} \\ H_1: \begin{cases} \boldsymbol{r}_l = \boldsymbol{a}\theta_l + \boldsymbol{c}_l, & l = 1,\dots,L \\ \boldsymbol{r}_l = \boldsymbol{c}_l, & l = L+1,\dots L+K \end{cases} \end{cases} \tag{3.27}$$

where $\boldsymbol{r}_l \in \mathbb{C}^{N\times1}, l = 1,\dots,L$ denotes the primary data, L denotes the range cells occupied by the target, $\boldsymbol{F} \in \mathbb{C}^{N\times(N-1)}$ is a known full-column-rank subspace matrix, $\boldsymbol{b}_l \in \mathbb{C}^{(N-1)\times1}$ is the unknown coordinate, $\boldsymbol{a} \in \mathbb{C}^{N\times1}$ is the target steering vector, θ_l is the unknown target coordinate, $\boldsymbol{r}_l \in \mathbb{C}^{N\times1}, l = L+1,\dots,L+K$ denote the training data, and K denotes the number of the training data. The noise $c_l = \sqrt{\tau_l} n_l \in \mathbb{C}^{N\times 1}$ satisfies the CG noise model [5, 9]. τ_l represents the random texture which obeys the gamma distribution, $\tau = \left[\tau_1,\ldots\tau_l,\ldots,\tau_L\right]$, $l = 1,\ldots,L$. The speckle n_l satisfies $n_l \sim \mathcal{CN}_N\left(0,\Sigma\right)$. The subspace matrix F satisfies: $\left\langle \Sigma^{-\frac{1}{2}} F \right\rangle = \left\langle \Sigma^{-\frac{1}{2}} a \right\rangle^{\perp}$.

3.2.2 Two-Step Generalized Likelihood Ratio Test

For the detection problem in Equation 3.27, the GLRT based upon the primary data with known noise covariance matrix is given by

$$\frac{\max_{\theta_1,\ldots,\theta_L} \int f\left(r_1,\ldots,r_L \mid \tau, H_1\right) f\left(\tau\right) d\tau}{\max_{b_1,\ldots,b_L} \int f\left(r_1,\ldots,r_L \mid \tau, H_0\right) f\left(\tau\right) d\tau} \underset{H_0}{\overset{H_1}{\gtrless}} \eta \tag{3.28}$$

where $f\left(\tau\right)$ denotes PDF of the texture, $f\left(r_1,\ldots,r_L\right)$ is the PDF of the primary data, η is the detection threshold. The PDF of the texture is [10]:

$$f\left(\tau_l\right) = \frac{1}{\beta_l^{v_l} \Gamma\left(v_l\right)} \tau_l^{v_l - 1} \exp\left(-\frac{\tau_l}{\beta_l}\right) \tag{3.29}$$

where β_l denotes the scale parameter, $\Gamma(\cdot)$ is the gamma function, v_l represents the shape parameter. The probability density functions of the primary data under hypotheses H_0 and H_1 are given by

$$f\left(r_1\ldots,r_L \mid \tau, H_0\right) = \prod_{l=1}^{L} \frac{1}{\pi^N \tau_l^N \det\left(\Sigma\right)} \times \exp\left[-\frac{1}{\tau_l}\left(r_l - F b_l\right)^H \Sigma^{-1} \left(r_l - F b_l\right)\right] \tag{3.30}$$

$$f\left(r_1\ldots,r_L \mid \tau, H_1\right) = \prod_{l=1}^{L} \frac{1}{\pi^N \tau_l^N \det\left(\Sigma\right)} \times \exp\left[-\frac{1}{\tau_l}\left(r_l - a\theta_l\right)^H \Sigma^{-1} \left(r_l - a\theta_l\right)\right] \tag{3.31}$$

We substitute the PDF of τ into the integration term of the denominator of Equation 3.28 and get the integration result

$$\max_{b_1,\ldots,b_L}\int f(r_1\ldots,r_L\mid\tau,H_0)f(\tau)d\tau$$

$$=\max_{b_1,\ldots,b_L}\prod_{l=1}^{L}\frac{2\left[\beta_l(r_l-Fb_l)^H\Sigma^{-1}(r_l-Fb_l)\right]^{\frac{\nu_l-N}{2}}}{\pi^N|\Sigma|\beta_l^{\nu_l}\Gamma(\nu_l)}\int\frac{\tau_l^{(\nu_l-1-N)}}{2}$$

$$\times\left[\beta_l(r_l-Fb_l)^H\Sigma^{-1}(r_l-Fb_l)\right]^{\frac{N-\nu_l}{2}}$$

$$\exp\left\{-\frac{\tau_l}{\beta_l}-\left[(r_l-Fb_l)^H\Sigma^{-1}(r_l-Fb_l)\right]/\tau_l\right\}d\tau_l$$

$$=\max_{b_1,\ldots,b_L}\prod_{l=1}^{L}\frac{2\left[\beta_l(r_l-Fb_l)^H\Sigma^{-1}(r_l-Fb_l)\right]^{\frac{\nu_l-N}{2}}}{\pi^N|\Sigma|\beta_l^{\nu_l}\Gamma(\nu_l)}$$

$$K_{N-\nu_l}\left[2\sqrt{(r_l-Fb_l)^H\Sigma^{-1}(r_l-Fb_l)/\beta_l}\right] \tag{3.32}$$

$$=\max_{b_1,\ldots,b_L}\prod_{l=1}^{L}\frac{\left[4(r_l-Fb_l)^H\Sigma^{-1}(r_l-Fb_l)/\beta_l\right]^{\frac{\nu_l-N}{2}}}{2^{\nu_l-1-N}\pi^N|\Sigma|\beta_l^N\Gamma(\nu_l)}$$

$$K_{N-\nu_l}\left[2\sqrt{\left[(r_l-Fb_l)^H\Sigma^{-1}(r_l-Fb_l)\right]/\beta_l}\right]$$

where $K_\alpha(\mu)$ denotes the modified Bessel function of the second kind. To estimate b_l, we set $y_l=2\sqrt{\left[(r_l-Fb_l)^H\Sigma^{-1}(r_l-Fb_l)\right]/\beta_l}$ and take the derivative of $y_l^{\nu_l-N}K_{N-\nu_l}(y_l)$ with respect to b_l. According to the property of the modified Bessel function of the second kind, we have

$$\begin{aligned}&d_{b_l}\left[y_l^{\nu_l-N}K_{N-\nu_l}(y_l)\right]\\&=-y_l^{\nu_l-N}K_{N-\nu_l+1}(y_l)d_{b_l}\left[2\sqrt{\left[(r_l-Fb_l)^H\Sigma^{-1}(r_l-Fb_l)\right]/\beta_l}\right]\end{aligned} \tag{3.33}$$

Since $y_l^{\nu_l-N}K_{N-\nu_l+1}(y_l)$ is not always be zero, the estimation of $\boldsymbol{b}_l$ is equivalent to setting $d_{\boldsymbol{b}_l}\left[2\sqrt{\left[(\boldsymbol{r}_l-\boldsymbol{F}\boldsymbol{b}_l)^H\Sigma^{-1}(\boldsymbol{r}_l-\boldsymbol{F}\boldsymbol{b}_l)\right]/\beta_l}\right]$ to zero. After some calculation, we get the estimation of $\boldsymbol{b}_l$

$$\hat{\boldsymbol{b}}_l=\left(\boldsymbol{F}^H\Sigma^{-1}\boldsymbol{F}\right)^{-1}\boldsymbol{F}^H\Sigma^{-1}\boldsymbol{r}_l \tag{3.34}$$

Then, the denominator of the GLRT is

$$\begin{aligned}
&\max_{\boldsymbol{b}_1,\ldots,\boldsymbol{b}_L}\int f(\boldsymbol{r}_1\ldots,\boldsymbol{r}_L\,|\,\tau,H_0)f(\tau)d\tau \\
&=\prod_{l=1}^{L}\frac{2\left[\beta_l\left(\boldsymbol{r}_l-\boldsymbol{F}\hat{\boldsymbol{b}}_l\right)^H\Sigma^{-1}\left(\boldsymbol{r}_l-\boldsymbol{F}\hat{\boldsymbol{b}}_l\right)\right]^{\frac{\nu_l-N}{2}}}{\pi^N|\Sigma|\beta_l^{\nu_l}\Gamma(\nu_l)}K_{N-\nu_l}\left[2\sqrt{\left(\boldsymbol{r}_l-\boldsymbol{F}\hat{\boldsymbol{b}}_l\right)^H\Sigma^{-1}\left(\boldsymbol{r}_l-\boldsymbol{F}\hat{\boldsymbol{b}}_l\right)/\beta_l}\right] \\
&\propto\prod_{l=1}^{L}\left(\boldsymbol{r}_l^H\Sigma^{-1}\boldsymbol{r}_l-\boldsymbol{r}_l^H\Sigma^{-\frac{1}{2}}\boldsymbol{P}_{\breve{\boldsymbol{F}}}\Sigma^{-\frac{1}{2}}\boldsymbol{r}_l\right)^{\frac{\nu_l-N}{2}}K_{N-\nu_l}\left[2\sqrt{\left(\boldsymbol{r}_l^H\Sigma^{-1}\boldsymbol{r}_l-\boldsymbol{r}_l^H\Sigma^{-\frac{1}{2}}\boldsymbol{P}_{\breve{\boldsymbol{F}}}\Sigma^{-\frac{1}{2}}\boldsymbol{r}_l\right)/\beta_l}\right]
\end{aligned} \tag{3.35}$$

where $\propto$ denotes "proportional to", $\breve{\boldsymbol{F}}=\Sigma^{-\frac{1}{2}}\boldsymbol{F}$, $\boldsymbol{P}_{\breve{\boldsymbol{F}}}=\breve{\boldsymbol{F}}\left(\breve{\boldsymbol{F}}^H\breve{\boldsymbol{F}}\right)^{-1}\breve{\boldsymbol{F}}^H$.

We can calculate the MLE of θ_l in a similar manner:

$$\hat{\theta}_l=\left(\boldsymbol{a}^H\Sigma^{-1}\boldsymbol{a}\right)^{-1}\boldsymbol{a}^H\Sigma^{-1}\boldsymbol{r}_l \tag{3.36}$$

The integration result under hypothesis H_1 is given by:

$$\begin{aligned}
&\max_{\theta_1,\ldots,\theta_L}\int f(\boldsymbol{r}_1,\ldots,\boldsymbol{r}_L\,|\,\tau,H_1)f(\tau)d\tau \\
&\propto\prod_{l=1}^{L}\left(\boldsymbol{r}_l^H\Sigma^{-1}\boldsymbol{r}_l-\boldsymbol{r}_l^H\Sigma^{-\frac{1}{2}}\boldsymbol{P}_{\breve{\boldsymbol{a}}}\Sigma^{-\frac{1}{2}}\boldsymbol{r}_l\right)^{\frac{\nu_l-N}{2}}K_{N-\nu_l}\left[2\sqrt{\left(\boldsymbol{r}_l^H\Sigma^{-1}\boldsymbol{r}_l-\boldsymbol{r}_l^H\Sigma^{-\frac{1}{2}}\boldsymbol{P}_{\breve{\boldsymbol{a}}}\Sigma^{-\frac{1}{2}}\boldsymbol{r}_l\right)/\beta_l}\right]
\end{aligned} \tag{3.37}$$

where $\breve{\boldsymbol{a}}=\Sigma^{-\frac{1}{2}}\boldsymbol{a}$, $\boldsymbol{P}_{\breve{\boldsymbol{a}}}=\Sigma^{-\frac{1}{2}}\boldsymbol{a}\left(\boldsymbol{a}^H\Sigma^{-1}\boldsymbol{a}\right)^{-1}\boldsymbol{a}^H\Sigma^{-\frac{1}{2}}$.

After some simplification, we find that the condition $\left\langle \Sigma^{-\frac{1}{2}}\boldsymbol{F}\right\rangle = \left\langle \Sigma^{-\frac{1}{2}}\boldsymbol{a}\right\rangle^{\perp}$ is equivalent to $\boldsymbol{I}_N - \boldsymbol{P}_{\breve{F}} = \boldsymbol{P}_{\breve{a}}$. The test statistic is obtained by plugging the above results into Equation 3.28:

$$\prod_{l=1}^{L}\left(\frac{\boldsymbol{r}_l^H\Sigma^{-1}\boldsymbol{r}_l - \boldsymbol{r}_l^H\Sigma^{-\frac{1}{2}}\boldsymbol{P}_{\breve{a}}\Sigma^{-\frac{1}{2}}\boldsymbol{r}_l}{\boldsymbol{r}_l^H\Sigma^{-\frac{1}{2}}\boldsymbol{P}_{\breve{a}}\Sigma^{-\frac{1}{2}}\boldsymbol{r}_l}\right)^{\frac{v_l-N}{2}}\frac{K_{N-v_l}\left[2\sqrt{\left(\boldsymbol{r}_l^H\Sigma^{-1}\boldsymbol{r}_l - \boldsymbol{r}_l^H\Sigma^{-\frac{1}{2}}\boldsymbol{P}_{\breve{a}}\Sigma^{-\frac{1}{2}}\boldsymbol{r}_l\right)/\beta_l}\right]}{K_{N-v_l}\left[2\sqrt{\boldsymbol{r}_l^H\Sigma^{-\frac{1}{2}}\boldsymbol{P}_{\breve{a}}\Sigma^{-\frac{1}{2}}\boldsymbol{r}_l/\beta_l}\right]}\underset{H_0}{\overset{H_1}{\gtrless}}\eta \tag{3.38}$$

The estimate of the noise covariance matrix is the fixed point covariance estimator $\hat{\Sigma}_{FP} = \frac{N}{K}\sum_{l=L+1}^{L+K}\frac{\boldsymbol{r}_l\boldsymbol{r}_l^H}{\boldsymbol{r}_l^H\hat{\Sigma}_{FP}^{-1}\boldsymbol{r}_l}$ in CG noise. In the second step, $\hat{\Sigma}_{FP}$ is substituted into test statistic to derive the adaptive detector. The final resulting selective GLRT in the CG noise (SE-GLRT) is

$$\prod_{l=1}^{L}\left(\frac{\boldsymbol{r}_l^H\hat{\Sigma}_{FP}^{-1}\boldsymbol{r}_l - \boldsymbol{r}_l^H\hat{\Sigma}_{FP}^{-\frac{1}{2}}\boldsymbol{P}_{\hat{a}}\hat{\Sigma}_{FP}^{-\frac{1}{2}}\boldsymbol{r}_l}{\boldsymbol{r}_l^H\hat{\Sigma}_{FP}^{-\frac{1}{2}}\boldsymbol{P}_{\hat{a}}\hat{\Sigma}_{FP}^{-\frac{1}{2}}\boldsymbol{r}_l}\right)^{\frac{v_l-N}{2}}$$

$$\frac{K_{N-v_l}\left[2\sqrt{\left(\boldsymbol{r}_l^H\hat{\Sigma}_{FP}^{-1}\boldsymbol{r}_l - \boldsymbol{r}_l^H\hat{\Sigma}_{FP}^{-\frac{1}{2}}\boldsymbol{P}_{\hat{a}}\hat{\Sigma}_{FP}^{-\frac{1}{2}}\boldsymbol{r}_l\right)/\beta_l}\right]}{K_{N-v_l}\left[2\sqrt{\boldsymbol{r}_l^H\hat{\Sigma}_{FP}^{-\frac{1}{2}}\boldsymbol{P}_{\hat{a}}\hat{\Sigma}_{FP}^{-\frac{1}{2}}\boldsymbol{r}_l/\beta_l}\right]}\underset{H_0}{\overset{H_1}{\gtrless}}\eta \tag{3.39}$$

where $\hat{\boldsymbol{a}} = \hat{\Sigma}_{FP}^{-\frac{1}{2}}\boldsymbol{a}$, $\boldsymbol{P}_{\hat{a}} = \hat{\boldsymbol{a}}\left(\hat{\boldsymbol{a}}^H\hat{\boldsymbol{a}}\right)^{-1}\hat{\boldsymbol{a}}^H$.

3.2.3 Maximum A Posteriori Generalized Likelihood Ratio Test

The MAP GLRT with known noise covariance matrix [11, 12] is:

$$\frac{\max_{\tau,\theta_1,\ldots,\theta_L}\prod_{l=1}^{L}f(\boldsymbol{r}_l\,|\,\tau_l,H_1)f(\tau_l)}{\max_{\tau,b_1,\ldots,b_L}\prod_{l=1}^{L}f(\boldsymbol{r}_l\,|\,\tau_l,H_0)f(\tau_l)}\underset{H_0}{\overset{H_1}{\gtrless}}\lambda \tag{3.40}$$

We obtain the estimation of $\boldsymbol{b}_l$ by minimizing the expression $\left(\boldsymbol{r}_l - \boldsymbol{F}\boldsymbol{b}_l\right)^H \Sigma^{-1}\left(\boldsymbol{r}_l - \boldsymbol{F}\boldsymbol{b}_l\right)$ over $\boldsymbol{b}_l$:

$$\hat{\boldsymbol{b}}_l = \left(\boldsymbol{F}^H\Sigma^{-1}\boldsymbol{F}\right)^{-1}\boldsymbol{F}^H\Sigma^{-1}\boldsymbol{z}_l \tag{3.41}$$

Then, we substitute the PDF of τ_l in Equation 3.29 and the estimation of $\boldsymbol{b}_l$ in Equation 3.41 into the denominator of the test statistic in Equation 3.40:

$$\prod_{l=1}^{L} f\left(\boldsymbol{r}_l \mid \tau_l, \hat{\boldsymbol{b}}_l, H_0\right) f\left(\tau_l\right) \propto \prod_{l=1}^{L} \tau_l^{(v_l-1-N)}$$

$$\exp\left\{-\tau_l/\beta_l - \left[\left(\boldsymbol{r}_l - \boldsymbol{F}\hat{\boldsymbol{b}}_l\right)^H \Sigma^{-1}\left(\boldsymbol{r}_l - \boldsymbol{F}\hat{\boldsymbol{b}}_l\right)\right]\Big/\tau_l\right\} \tag{3.42}$$

We set $\boldsymbol{T}_{l,0} = \left(\boldsymbol{r}_l - \boldsymbol{F}\hat{\boldsymbol{b}}_l\right)^H \Sigma^{-1}\left(\boldsymbol{r}_l - \boldsymbol{F}\hat{\boldsymbol{b}}_l\right)$. By using the condition that $\boldsymbol{I}_N - \boldsymbol{P}_{\breve{\boldsymbol{F}}} = \boldsymbol{P}_{\breve{\boldsymbol{a}}}$, $\boldsymbol{T}_{l,0}$ can also be written as $\boldsymbol{T}_{l,0} = \boldsymbol{r}_l^H\Sigma^{-1}\boldsymbol{r}_l - \boldsymbol{r}_l^H\Sigma^{-\frac{1}{2}}\boldsymbol{P}_{\breve{\boldsymbol{F}}}\Sigma^{-\frac{1}{2}}\boldsymbol{r}_l = \boldsymbol{r}_l^H\Sigma^{-\frac{1}{2}}\boldsymbol{P}_{\breve{\boldsymbol{a}}}\Sigma^{-\frac{1}{2}}\boldsymbol{r}_l$. In order to maximize Equation 3.42 over τ_l, we differentiate Equation 3.42 with respect to τ_l and set the derivative to zero. The result is given by:

$$\hat{\tau}_{l,0} = \frac{\beta_l\left[\left(v_l - 1 - N\right) + \sqrt{\left(v_l - 1 - N\right)^2 + 4\boldsymbol{T}_{l,0}/\beta_l}\right]}{2} \tag{3.43}$$

Inserting (3.43) into (3.42) results in

$$\prod_{l=1}^{L} f\left(\boldsymbol{r}_l \mid \hat{\tau}_{l,0}, \hat{\boldsymbol{b}}_l, H_0\right) f\left(\tau_l\right) \propto \prod_{l=1}^{L} \hat{\tau}_{l,0}^{(v_l-1-N)} \exp\left(-\hat{\tau}_{l,0}/\beta_l - \boldsymbol{T}_{l,0}/\hat{\tau}_{l,0}\right) \tag{3.44}$$

Similarly, we can get the estimation of τ_l under H_1:

$$\hat{\tau}_{l,1} = \frac{\beta_l}{2}\left[\left(v_l - 1 - N\right) + \sqrt{\left(v_l - 1 - N\right)^2 + 4\boldsymbol{T}_{l,1}/\beta_l}\right] \tag{3.45}$$

where $\boldsymbol{T}_{l,1} = \left(\boldsymbol{r}_l - \boldsymbol{a}\hat{\theta}_l\right)^H \Sigma^{-1}\left(\boldsymbol{r}_l - \boldsymbol{a}\hat{\theta}_l\right) = \boldsymbol{r}_l^H\Sigma^{-1}\boldsymbol{r}_l - \boldsymbol{r}_l^H\Sigma^{-\frac{1}{2}}\boldsymbol{P}_{\breve{\boldsymbol{a}}}\Sigma^{-\frac{1}{2}}\boldsymbol{r}_l$.

Then, we substitute Equations 3.44 and 3.45 into the test statistic Equation 3.40 and get the MAP GLRT with a known noise covariance matrix:

$$\prod_{l=1}^{L}\frac{\hat{\tau}_{l,1}^{(v_l-1-N)}\exp\left\{-\hat{\tau}_{l,1}/\beta_l-\left[\boldsymbol{r}_l^H\Sigma^{-\frac{1}{2}}\left(\boldsymbol{I}_N-\boldsymbol{P}_{\breve{a}}\right)\Sigma^{-\frac{1}{2}}\boldsymbol{r}_l\right]/\hat{\tau}_{l,1}\right\}}{\hat{\tau}_{l,0}^{(v_l-1-N)}\exp\left(-\hat{\tau}_{l,0}/\beta_l-\boldsymbol{r}_l^H\Sigma^{-\frac{1}{2}}\boldsymbol{P}_{\breve{a}}\Sigma^{-\frac{1}{2}}\boldsymbol{r}_l/\hat{\tau}_{l,0}\right)}\underset{H_0}{\overset{H_1}{\gtrless}}\ln\lambda \tag{3.46}$$

Finally, $\hat{\Sigma}_{FP}$ is substituted into Equation 3.46 to replace the known covariance matrix and the final adaptive selective MAP in the CG noise (SE-MAP) is then obtained. The SE-GLRT and the SE-MAP have the CFAR property with respect to Σ and the proof is given in the Appendix 3.A.

3.2.4 Numerical Examples

The detection performance of the SE-GLRT and the SE-MAP is assessed through Monte Carlo simulations. We set $N=8$, $\Sigma(i,j)=\rho^{|i-j|}, i,j=1,\ldots,N$, $\rho=0.9$, $P_{fa}=10^{-3}$, $K=16, L=3, \boldsymbol{a}=[1,\ldots,1]^T/\sqrt{N}$, $v_1=\cdots=v_{L+K}=v=4$, $\beta_1=\cdots=\beta_{L+K}=1/v$. The iteration number is three to compute $\hat{\Sigma}_{FP}$. To obtain probabilities of detection and thresholds, 10^5 independent Monte Carlo simulation trials are conducted. The SNR is

$$\text{SNR}=\sum_{l=1}^{L}|\theta_l|^2\boldsymbol{a}^H\Sigma^{-1}\boldsymbol{a} \tag{3.47}$$

3.2.4.1 Mismatched Detection Performance Analysis

The mismatched detection performance of the SE-GLRT and SE-MAP is analyzed. The mismatch angle and the SNR are defined according to the actual target steering vector $\boldsymbol{a}_0$ and presumed one $\boldsymbol{a}$ [13, 14]:

$$\cos^2\phi=\frac{\left|\left(\boldsymbol{a}^H\Sigma^{-1}\boldsymbol{a}_0\right)\right|^2}{\left(\boldsymbol{a}^H\Sigma^{-1}\boldsymbol{a}\right)\left(\boldsymbol{a}_0^H\Sigma^{-1}\boldsymbol{a}_0\right)},\ \text{SNR}=\sum_{l=1}^{L}|\theta_l|^2\boldsymbol{a}_0^H\Sigma^{-1}\boldsymbol{a}_0.$$

For comparison, the GLRT in the CG noise [15], which is also designed for the distributed targets in the CG noise with gamma texture, is given. The test statistic of the GLRT in CG noise is given by:

$$\prod_{l=1}^{L}\left(\frac{\boldsymbol{r}_l^H\hat{\Sigma}_{FP}^{-1}\boldsymbol{r}_l-\boldsymbol{r}_l^H\hat{\Sigma}_{FP}^{-\frac{1}{2}}\boldsymbol{P}_{\hat{a}}\hat{\Sigma}_{FP}^{-\frac{1}{2}}\boldsymbol{r}_l}{\boldsymbol{r}_l^H\hat{\Sigma}_{FP}^{-1}\boldsymbol{r}_l}\right)^{\frac{v_l-N}{2}}$$

$$\frac{K_{N-v_l}\left(2\sqrt{\left(\boldsymbol{r}_l^H\hat{\Sigma}_{FP}^{-1}\boldsymbol{r}_l-\boldsymbol{r}_l^H\hat{\Sigma}_{FP}^{-\frac{1}{2}}\boldsymbol{P}_{\hat{a}}\hat{\Sigma}_{FP}^{-\frac{1}{2}}\boldsymbol{r}_l\right)/\beta_l}\right)}{K_{N-v_l}\left(2\sqrt{\boldsymbol{r}_l^H\hat{\Sigma}_{FP}^{-1}\boldsymbol{r}_l/\beta_l}\right)}\underset{H_0}{\overset{H_1}{\gtrless}}\eta_{GLRT} \tag{3.48}$$

In Figure 3.3, contours of const P_d of the SE-GLRT, SE-MAP, and the GLRT in the CG noise are plotted. The curves show that the SE-GLRT and the SE-MAP have better discrimination ability for mismatch signals. Meanwhile, the mismatched detection performance of the SE-GLRT and the SE-MAP is similar.

3.2.4.2 Matched Detection Performance Analysis

The matched detection performance of the SE-GLRT and the SE-MAP is evaluated. In Figure 3.4, probabilities of detection of the SE-GLRT and SE-MAP are plotted against SNR. Figure 3.4 shows that the SE-GLRT and the SE-MAP suffer some matched detection performance degradation compared with the GLRT in the CG noise. The performance loss of the SE-GLRT and the SE-MAP with respect to the GLRT-CG is less than 1 dB. Meanwhile, the SE-GLRT and the SE-MAP have similar matched detection performance. Thus, the SE-GLRT and SE-MAP are more selective to the mismatched signals at the price of slightmatched detection performance degradation.

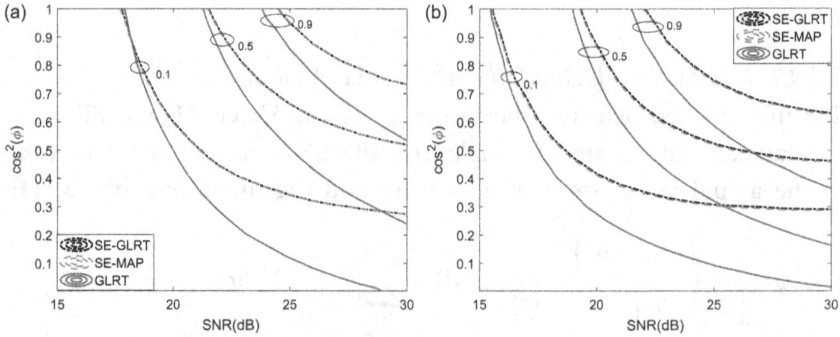

FIGURE 3.3 Contours of const P_d for the SE-GLRT, SE-MAP, and GLRT (a) $K=20$; (b) $K=30$.

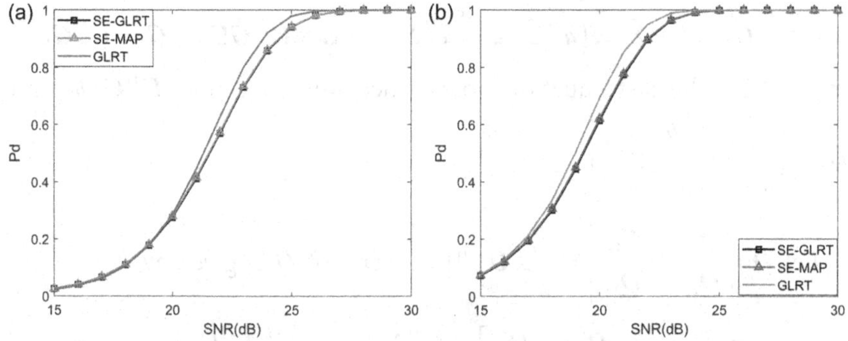

FIGURE 3.4 The probability of detection versus SNR. (a) $K=20$; (b) $K=30$.

3.3 CONCLUSION

In this chapter, we considered the problem of signal detection when a signal mismatch occurs. The unwanted signals were added under the null hypothesis to enhance the selectivity and four selective detectors were derived. Specifically, the SE-1SGLRT and the SE-2SGLRT were derived in the PHE wherein the training data have the same noise covariance matrix with the primary data up to an unknown scaling factor. The SE-GLRT and the SE-MAP were designed in the CG noise wherein the texture follows the gamma distribution. The SE-1SGLRT, the SE-2SGLRT, the SE-GLRT, and the SE-MAP have stronger mismatch signal discrimination capacities than their counterparts.

APPENDIX 3.A THE CFAR PROPERTY ANALYSIS

To demonstrate the CFAR property of the SE-GLRT and the SE-MAP, we define $\tilde{\boldsymbol{n}}_l = \boldsymbol{\Sigma}^{-\frac{1}{2}}\boldsymbol{n}_l, \tilde{\boldsymbol{a}} = \boldsymbol{\Sigma}^{-\frac{1}{2}}\boldsymbol{a}$, $\tilde{\boldsymbol{\Sigma}} = \boldsymbol{\Sigma}^{-\frac{1}{2}}\breve{\boldsymbol{\Sigma}}_{FP}\boldsymbol{\Sigma}^{-\frac{1}{2}}$. There exists a unitary matrix $\boldsymbol{U}=\left[\boldsymbol{U}_1,\boldsymbol{U}_2\right]$ which satisfies $\boldsymbol{U}_1=\tilde{\boldsymbol{a}}\left(\tilde{\boldsymbol{a}}^H\tilde{\boldsymbol{a}}\right)^{-\frac{1}{2}}$, $\boldsymbol{U}_2^H\boldsymbol{U}_1=\boldsymbol{0}_{(N-1)\times 1}$, $\boldsymbol{U}_2^H\boldsymbol{U}_2=\boldsymbol{0}_{(N-1)\times(N-1)}$. Let $\bar{\boldsymbol{n}}_l = \boldsymbol{U}^H\tilde{\boldsymbol{n}}_l$, $\bar{\boldsymbol{a}} = \boldsymbol{U}^H\tilde{\boldsymbol{a}}$. The expression $\boldsymbol{r}_l^H\hat{\boldsymbol{\Sigma}}_{FP}^{-1}\boldsymbol{r}_l - \boldsymbol{r}_l^H\hat{\boldsymbol{\Sigma}}_{FP}^{-\frac{1}{2}}\boldsymbol{P}_{\hat{\boldsymbol{a}}}\hat{\boldsymbol{\Sigma}}_{FP}^{-\frac{1}{2}}\boldsymbol{r}_l$ is simplified as

$$
\begin{aligned}
&\boldsymbol{r}_l^H\hat{\boldsymbol{\Sigma}}_{FP}^{-1}\boldsymbol{r}_l - \boldsymbol{r}_l^H\hat{\boldsymbol{\Sigma}}_{FP}^{-\frac{1}{2}}\boldsymbol{P}_{\hat{\boldsymbol{a}}}\hat{\boldsymbol{\Sigma}}_{FP}^{-\frac{1}{2}}\boldsymbol{r}_l \\
&=\tau_l\boldsymbol{n}_l^H\left[\hat{\boldsymbol{\Sigma}}_{FP}^{-1} - \hat{\boldsymbol{\Sigma}}_{FP}^{-1}\boldsymbol{a}\left(\boldsymbol{a}^H\hat{\boldsymbol{\Sigma}}_{FP}^{-1}\boldsymbol{a}\right)^{-1}\boldsymbol{a}^H\hat{\boldsymbol{\Sigma}}_{FP}^{-1}\right]\boldsymbol{n}_l \qquad (3.49)\\
&=\tau_l\tilde{\boldsymbol{n}}_l^H\tilde{\boldsymbol{G}}\tilde{\boldsymbol{n}}_l=\tau_l\bar{\boldsymbol{n}}_l^H\bar{\boldsymbol{G}}\bar{\boldsymbol{n}}_l=\tau_l\bar{\boldsymbol{n}}_{2l}^H\bar{\boldsymbol{G}}_{22}\bar{\boldsymbol{n}}_{2l}
\end{aligned}
$$

where $\tilde{G}=\tilde{\Sigma}^{-1}-\tilde{\Sigma}^{-1}\tilde{a}\left(\tilde{a}^H\tilde{\Sigma}^{-1}\tilde{a}\right)^{-1}\tilde{a}^H\tilde{\Sigma}^{-1}$, $\bar{G}=U^H\tilde{G}U$, $\bar{G}_{22}=U_2^H\tilde{G}U_2$, $\bar{n}_{2l}=U_2^H\tilde{n}_l$. The last equation holds since we prove that $U_1^H\tilde{G}=0$ and $\bar{G}=\begin{bmatrix} 0_{1\times 1} & 0_{1\times(N-1)} \\ 0_{(N-1)\times 1} & \bar{G}_{22}\end{bmatrix}$.

Let $Q=\begin{bmatrix} Q_{11} & Q_{12} \\ Q_{21} & Q_{22}\end{bmatrix}=U^H\tilde{\Sigma}U$. The inverse of Q is given by

$$Q^{-1}=\begin{bmatrix} Q'_{11} & Q'_{12} \\ Q'_{21} & Q'_{22}\end{bmatrix}=\begin{bmatrix} U_1^H\tilde{\Sigma}^{-1}U_1 & U_1^H\tilde{\Sigma}^{-1}U_2 \\ U_2^H\tilde{\Sigma}^{-1}U_1 & U_2^H\tilde{\Sigma}^{-1}U_2\end{bmatrix} \tag{3.50}$$

Utilizing the partitioned matrix inversion theorem, we rewrite $\bar{G}_{22}$ as

$$\begin{aligned}\bar{G}_{22}&=U_2^H\tilde{\Sigma}^{-1}U_2-U_2^H\tilde{\Sigma}^{-1}U_1\left(U_1^H\tilde{\Sigma}^{-1}U_1\right)^{-1}U_1^H\tilde{\Sigma}^{-1}U_2 \\ &=Q'_{22}-Q'_{21}Q'^{-1}_{11}Q'_{12}=Q_{22}^{-1}\end{aligned} \tag{3.51}$$

It has been shown in [16] and [17] that $\tilde{\Sigma}=\Sigma^{-\frac{1}{2}}\hat{\Sigma}_{FP}\Sigma^{-\frac{1}{2}}$ is the unique fixed point estimator (up to a scale factor) of the identity matrix. Thus, Q_{22} is independent of the noise covariance matrix Σ. Expression $r_l^H\hat{\Sigma}_{FP}^{-1}r_l-r_l^H\hat{\Sigma}_{FP}^{-\frac{1}{2}}P_{\hat{a}}\hat{\Sigma}_{FP}^{-\frac{1}{2}}r_l$ is independent of Σ since $\bar{n}_{2l}$, Q_{22}, τ_l are independent of Σ.

Next, the CFAR property of the expression $r_l^H\hat{\Sigma}_{FP}^{-\frac{1}{2}}P_{\hat{a}}\hat{\Sigma}_{FP}^{-\frac{1}{2}}r_l$ is analyzed:

$$\begin{aligned}&r_l^H\hat{\Sigma}_{FP}^{-\frac{1}{2}}P_{\hat{a}}\hat{\Sigma}_{FP}^{-\frac{1}{2}}r_l \\ &=\tau_l\tilde{n}_l^H\tilde{\Sigma}^{-1}U_1\left(U_1^H\tilde{\Sigma}^{-1}U_1\right)^{-1}U_1^H\tilde{\Sigma}^{-1}\tilde{n}_l \\ &=\tau_l\bar{n}_l^HQ^{-1}P\left(P^HQ^{-1}P\right)^{-1}P^HQ^{-1}\bar{n}_l\end{aligned} \tag{3.52}$$

where $P=U^HU_1=\begin{bmatrix}1 \\ 0_{(N-1)\times 1}\end{bmatrix}$. We substitute the definitions of $\bar{n}_l$, Q^{-1} and P into the above equation and get

$$\begin{aligned}P^HQ^{-1}\bar{n}_l&=\begin{bmatrix}1 & 0_{1\times(N-1)}\end{bmatrix}\begin{bmatrix} Q'_{11} & Q'_{12} \\ Q'_{21} & Q'_{22}\end{bmatrix}\begin{bmatrix}\bar{n}_{1l} \\ \bar{n}_{2l}\end{bmatrix} \\ &=Q'_{11}\bar{n}_{1l}+Q'_{12}\bar{n}_{2l}=Q'_{11}\left(\bar{n}_{1l}-Q_{12}Q_{22}^{-1}\bar{n}_{2l}\right)\end{aligned} \tag{3.53}$$

And $\boldsymbol{P}^H\boldsymbol{Q}^{-1}\boldsymbol{P} = \boldsymbol{Q}'_{11}$. Then, $\boldsymbol{r}_l^H\hat{\Sigma}_{FP}^{-\frac{1}{2}}\boldsymbol{P}_{\hat{a}}\hat{\Sigma}_{FP}^{-\frac{1}{2}}\boldsymbol{r}_l$ is simplified as

$$\boldsymbol{r}_l^H\hat{\Sigma}_{FP}^{-\frac{1}{2}}\boldsymbol{P}_{\hat{a}}\hat{\Sigma}_{FP}^{-\frac{1}{2}}\boldsymbol{r}_l = \tau_l\left(\bar{\boldsymbol{n}}_{1l} - \boldsymbol{Q}_{12}\boldsymbol{Q}_{22}^{-1}\bar{\boldsymbol{n}}_{2l}\right)^H \boldsymbol{Q}'_{11}\left(\bar{\boldsymbol{n}}_{1l} - \boldsymbol{Q}_{12}\boldsymbol{Q}_{22}^{-1}\bar{\boldsymbol{n}}_{2l}\right) \quad (3.54)$$

Since $\bar{\boldsymbol{n}}_{1l}$, $\bar{\boldsymbol{n}}_{2l}$, $\boldsymbol{Q}$, τ_l are independent of the noise covariance matrix Σ, the expression $\boldsymbol{r}_l^H\hat{\Sigma}_{FP}^{-\frac{1}{2}}\boldsymbol{P}_{\hat{a}}\hat{\Sigma}_{FP}^{-\frac{1}{2}}\boldsymbol{r}_l$ is independent of Σ.

From the test statistics of the SE-GLRT in Equation 3.39 and the SE-MAP in Equation 3.46, we can see that the test statistics of the two detectors are functions of expressions $\boldsymbol{r}_l^H\hat{\Sigma}_{FP}^{-1}\boldsymbol{r}_l - \boldsymbol{r}_l^H\hat{\Sigma}_{FP}^{-\frac{1}{2}}\boldsymbol{P}_{\hat{a}}\hat{\Sigma}_{FP}^{-\frac{1}{2}}\boldsymbol{r}_l$ and $\boldsymbol{r}_l^H\hat{\Sigma}_{FP}^{-\frac{1}{2}}\boldsymbol{P}_{\hat{a}}\hat{\Sigma}_{FP}^{-\frac{1}{2}}\boldsymbol{r}_l$. Thus, the SE-GLRT and the SE-MAP are CFAR with respect to the Σ.

REFERENCES

1. S. Kraut and L. L. Scharf, "The CFAR Adaptive Subspace is a Scale-Invariant GLRT," *IEEE Transactions on Signal Processing*, vol. 47, no. 9, pp. 2538–2541, 1999.
2. N. B. Pulsone and C. M. Rader, "Adaptive Beamformer Orthogonal Rejection Test," *IEEE Transactions on Signal Processing*, vol. 49, no. 3, pp. 521–529, 2001.
3. F. Bandiera, O. Besson, and G. Ricci, "An Abort-Like Detector with Improved Mismatched Signals Rejection Capabilities," *IEEE Transactions on Signal Processing*, vol. 56, pp. 14–25, 2008.
4. G. A. Fabrizio, A. Farina, and M. D. Turley, "Spatial Adaptive Subspace Detection in OTH Radar," *IEEE Transactions on Aerospace and Electronic Systems*, vol. 39, no. 4, pp. 1407–1428, 2003.
5. F. Bandiera, O. Besson, A. Coluccia, and G. Ricci, "ABORT-Like Detectors: A Bayesian Approach," *IEEE Transactions on Signal Processing*, vol. 63, no. 19, pp. 5274–5284, 2015.
6. D. Orlando and G. Ricci, "A Rao Test with Enhanced Selectivity Properties in Homogeneous Scenarios," *IEEE Transactions on Signal Processing*, vol. 58, no. 10, pp. 5385–5390, 2010.
7. E. Conte, A. De Maio, and G. Ricci, "GLRT-Based Adaptive Detection Algorithms for Range-Spread Targets," *IEEE Transactions on Signal Processing*, vol. 49, no. 7, pp. 1336–1348, 2001.
8. J. Carretero-Moya, J. Gismero-Menoyo, Á. Blanco-del-Campo, and A. Asensio-Lopez, "Statistical Analysis of a High-Resolution Sea-Clutter Database," *IEEE Transactions on Geoscience and Remote Sensing*, vol. 48, no. 4, pp. 2024–2037, 2010.
9. K. Gerlach, "Spatially Distributed Target Detection in Non-Gaussian Clutter," *IEEE Transactions on Aerospace and Electronic Systems*, vol. 35, no. 3, pp. 926–934, 1999.

10. L. Rosenberg and S. Bocquet, "Non-Coherent Radar Detection Performance in Medium Grazing Angle X-Band Sea Clutter," *IEEE Transactions on Aerospace and Electronic Systems*, vol. 53, no. 2, pp. 669–682, 2017.
11. X. Shang, H. Song, Y. Wang, C. Hao, and C. Lei, "Adaptive Detection of Distributed Targets in Compound-Gaussian Clutter with Inverse Gamma Texture," *Digital Signal Processing*, vol. 22, no. 6, pp. 1024–1030, December 2012.
12. S. Xu, J. Xue, and P. Shui, "Adaptive Detection of Range-Spread Targets in Compound Gaussian Clutter with The Square Root of Inverse Gaussian Texture," *Digital Signal Processing*, vol. 56, pp. 132–139, 2016.
13. R. S. Raghavan, "Analysis of Steering Vector Mismatch on Adaptive Noncoherent Integration," *IEEE Transactions on Aerospace and Electronic Systems*, vol. 49, no. 4, pp. 2496–2508, 2013.
14. W. Liu, J. Liu, L. Huang, Z. Yang, H. Yang, and Y. Wang, "Distributed Target Detectors with Capabilities of Mismatched Subspace Signal Rejection," *IEEE Transactions on Aerospace and Electronic Systems*, vol. 53, no. 2, pp. 888–900, 2017.
15. X. Shuai, L. Kong, and J. Yang, "Performance Analysis of GLRT-Based Adaptive Detector for Distributed Targets in Compound-Gaussian Clutter," *Signal Processing*, vol. 90, no. 1, pp. 16–23, 2010.
16. Y. C. Gao, G. S. Liao, S. Q. Zhu, and D. Yang, "A Persymmetric GLRT for Adaptive Detection in Compound-Gaussian Clutter with Random Texture," *IEEE Signal Processing Letters*, vol. 20, no. 6, pp. 615–618, 2013.
17. F. Pascal, P. Forster, J. -P. Ovarlez, and P. Larzabal, "Performance Analysis of Covariance Matrix Estimates in Impulsive Noise," *IEEE Transactions on Signal Processing*, vol. 56, no. 6, pp. 2206–2217, 2008.

CHAPTER 4

Adaptive Robust Detector in the Presence of Signal Mismatch

In Chapter 3, the design of adaptive detectors with enhanced selectivity is discussed. However, in some scenarios such as radars in scan mode, a detector that is robust to the mismatched signals is preferred. The robust detectors still have a high PD when the signal mismatch occurs [1–3]. As to the conventional adaptive detectors which are derived in the matched signal cases, the AMF exhibits better robustness than Kelly's GLRT. To improve the robustness of the adaptive detectors, three methods can be used. One method is to exploit the subspace signal model. In [4], the target signal is assumed to lie in a known subspace, and a GLRT detector is derived in Gaussian noise with persymmetric covariance matrix. The resulting persymmetric GLRT detector exhibits stronger robustness to the mismatched signals than Cai-Wang's GLRT detector derived in [5]. The second method is to constrain the target steering vector in a cone. In [6], the real parts and imaginary parts of the target steering vector are assumed to belong to the union of two convex cones, and three GLRT-based adaptive detectors are derived. The three detectors perform better than Kelly's GLRT, AMF, and ABORT in mismatched signal cases. In [1], the cone is defined by the target steering vector, and a semidefinite programming problem is solved to derive the GLRT. The resulting GLRT detector is more robust than the generalized adaptive subspace detector

DOI: 10.1201/9781003477907-4

derived in [7] based on the two-step GLRT. Another method is to modify the binary hypothesis by introducing a random component under the alternative hypothesis. In [8], a random signal is introduced under the alternative hypothesis, and GLRT-based detectors are designed to improve the detection performance in mismatched signal cases. In this chapter, the problem of detecting a point-like target in the Gaussian noise with an unknown covariance matrix is discussed. To enhance the robustness, the gradient test is derived. The CFAR property of the gradient detector is also analyzed.

4.1 PROBLEM FORMULATION

The received radar echoes are denoted by $\boldsymbol{x}_0 \in \mathbb{C}^{N\times1}$, where N is the number of coherent pulses or the product of pulse numbers and antenna array elements [8, 9]. We formulate the problem of target detection as binary hypothesis testing. Under the alternative hypothesis, the echoes include the noise $\boldsymbol{n}_0$ and useful target signal $\alpha\boldsymbol{s}$. Under null hypothesis, the echoes include only the noise $\boldsymbol{n}_0$. Here, $\boldsymbol{n}_0 \in \mathbb{C}^{N\times1}$ follows the complex Gaussian distribution with zero-mean and covariance matrix $\boldsymbol{R}$, α denotes the target amplitude, $\boldsymbol{s} \in \mathbb{C}^{N\times1}$ is the steering vector of the target. To make the alternative hypothesis more plausible, a random component $\boldsymbol{\omega}$ is added to the received data under the alternative hypothesis. Doing so, the designed detector will be more inclined to declare a detection than the case where the received data contain only the noise and possible target signal. The detection problem is equivalent to deciding which of the two hypotheses holds:

$$\begin{cases} H_0: \begin{cases} \boldsymbol{x}_0 = \boldsymbol{n}_0, \\ \boldsymbol{x}_k = \boldsymbol{n}_k, k = 1,\dots,K \end{cases} \\ H_1: \begin{cases} \boldsymbol{x}_0 = \alpha\boldsymbol{s} + \boldsymbol{\omega} + \boldsymbol{n}_0, \\ \boldsymbol{x}_k = \boldsymbol{n}_k, k = 1,\dots,K \end{cases} \end{cases} \tag{4.1}$$

where $\boldsymbol{x}_0$ denotes data under test or primary data, $\boldsymbol{X}_k = \left[\boldsymbol{x}_1,\dots\boldsymbol{x}_k,\dots,\boldsymbol{x}_K\right], k = 1,\dots,K$ denotes training data. The random perturbation $\boldsymbol{\omega}$ obeys zero-mean complex Gaussian distribution with covariance matrix $\delta\boldsymbol{R}$.

Then, the distributions of the received data under the two hypotheses are

$$\begin{cases} H_0 : \begin{cases} \boldsymbol{x}_0 \sim \mathcal{CN}_N(\mathbf{0}, \boldsymbol{R}), \\ \boldsymbol{x}_k \sim \mathcal{CN}_N(\mathbf{0}, \boldsymbol{R}), k = 1, \ldots, K \end{cases} \\ H_1 : \begin{cases} \boldsymbol{x}_0 \sim \mathcal{CN}_N(\alpha \boldsymbol{s}, (1+\delta)\boldsymbol{R}), \\ \boldsymbol{x}_k \sim \mathcal{CN}_N(\mathbf{0}, \boldsymbol{R}), k = 1, \ldots, K \end{cases} \end{cases} \tag{4.2}$$

where δ is a real constant greater than 0.

According to the distribution of the received data, we calculate the joint PDF of the training data $\boldsymbol{X}_k$ and the primary data $\boldsymbol{x}_0$ under H_μ as

$$f_\mu = \frac{1}{\pi^{N(1+K)}(1+\delta)^N \det^{(K+1)}(\boldsymbol{R})}$$

$$\exp\left\{-\mathrm{tr}\left[\boldsymbol{R}^{-1}\left(\boldsymbol{S} + \frac{1}{(1+\mu\delta)}(\boldsymbol{x}_0 - \mu\alpha\boldsymbol{s})(\boldsymbol{x}_0 - \mu\alpha\boldsymbol{s})^H\right)\right]\right\} \tag{4.3}$$

where $\mu = 0,1$, $\boldsymbol{S} = \sum_{k=1}^{K} \boldsymbol{x}_k \boldsymbol{x}_k^H$.

4.2 ADAPTIVE ROBUST GRADIENT DETECTOR

We set $\Theta_r = \left[\alpha^T, \alpha^H, \delta\right]^T$, $\Theta_s = \mathrm{vec}(\boldsymbol{R})$, $\Theta = \left[\Theta_r^T, \Theta_s^T\right]^T$, where Θ is an (N^2+3)-dimensional column vector. We resort to a one-step complex parameter gradient test to design an adaptive detector.

The complex parameter gradient test is

$$T_{Gradient} = \left.\frac{\partial \ln f_1}{\partial \Theta_r^T}\right|_{\Theta=\hat{\Theta}_0} \left(\hat{\Theta}_{r1} - \Theta_{r0}\right) \underset{H_0}{\overset{H_1}{\gtrless}} \eta \tag{4.4}$$

where $\hat{\Theta}_0$ is the MLE of Θ under H_0, Θ_{r0} is the value of Θ_r under H_0, $\hat{\Theta}_{r1}$ is the MLE of Θ_r under H_1.

We substitute Θ_r into the test statistic (4.4) and get

$$\frac{\partial \ln f_1}{\partial \Theta_r^T} = \left(\frac{\partial \ln f_1}{\partial \Theta_r^*}\right)^H = \begin{bmatrix} \dfrac{\partial \ln f_1}{\partial \alpha^*} \\ \dfrac{\partial \ln f_1}{\partial \alpha} \\ \dfrac{\partial \ln f_1}{\partial \delta} \end{bmatrix}^H \tag{4.5}$$

According to differential theory of complex matrices, we take the partial differential of the logarithm of the joint PDF (4.3) under H_1 with respect to Θ_r^* and have the following results

$$\frac{\partial \ln f_1}{\partial \alpha^*} = \frac{1}{1+\delta}\left(\boldsymbol{s}^H \boldsymbol{R}^{-1}\boldsymbol{x}_0 - \alpha \boldsymbol{s}^H \boldsymbol{R}^{-1}\boldsymbol{s}\right) \tag{4.6}$$

$$\frac{\partial \ln f_1}{\partial \alpha} = \frac{1}{1+\delta}\left(\boldsymbol{x}_0^H \boldsymbol{R}^{-1}\boldsymbol{s} - \alpha^H \boldsymbol{s}^H \boldsymbol{R}^{-1}\boldsymbol{s}\right)^T \tag{4.7}$$

$$\frac{\partial \ln f_1}{\partial \delta} = -\frac{N}{1+\delta} + \frac{1}{(1+\delta)^2}\left[(\boldsymbol{x}_0 - \alpha\boldsymbol{s})^H \boldsymbol{R}^{-1}(\boldsymbol{x}_0 - \alpha\boldsymbol{s})\right] \tag{4.8}$$

From the test statistic in Equation 4.4, it can be seen that the MLE of Θ under H_0 and the MLE of Θ_r under H_1 are also needed. After some calculation, we obtain

$$\begin{aligned} d_{\boldsymbol{R}} \ln f_1 &= -(K+1)\operatorname{tr}\left(\boldsymbol{R}^{-1} d\boldsymbol{R}\right) \\ &\quad + \operatorname{tr}\left\{\left[\boldsymbol{S} + (\boldsymbol{x}_0 - \alpha\boldsymbol{s})(\boldsymbol{x}_0 - \alpha\boldsymbol{s})^H/(1+\delta)\right]\boldsymbol{R}^{-1} d\boldsymbol{R}\boldsymbol{R}^{-1}\right\} \end{aligned} \tag{4.9}$$

$$\hat{\boldsymbol{R}}_1 = \left[\boldsymbol{S} + \frac{1}{(1+\delta)}(\boldsymbol{x}_0 - \alpha\boldsymbol{s})(\boldsymbol{x}_0 - \alpha\boldsymbol{s})^H\right] \Big/ (K+1) \tag{4.10}$$

$$\hat{\boldsymbol{R}}_0 = \left(\boldsymbol{S} + \boldsymbol{x}_0\boldsymbol{x}_0^H\right)/(K+1) \tag{4.11}$$

$$\hat{\Theta}_0 = \left[0_{1\times 3}, \operatorname{vec}^T\left(\hat{\boldsymbol{R}}_0\right)\right]^T \tag{4.12}$$

$$\begin{aligned}\hat{\alpha} &= \arg\min_{\alpha} \left| \boldsymbol{S} + (\boldsymbol{x}_0 - \alpha\boldsymbol{s})(\boldsymbol{x}_0 - \alpha\boldsymbol{s})^H / (1+\delta) \right| \\ &= \arg\min_{\alpha} (1+\delta)^{-1} |\boldsymbol{S}| \left[(1+\delta) + (\boldsymbol{x}_0 - \alpha\boldsymbol{s})^H \boldsymbol{S}^{-1} (\boldsymbol{x}_0 - \alpha\boldsymbol{s}) \right] \\ &= \frac{\boldsymbol{s}^H \boldsymbol{S}^{-1} \boldsymbol{x}_0}{\boldsymbol{s}^H \boldsymbol{S}^{-1} \boldsymbol{s}}\end{aligned} \tag{4.13}$$

$$\begin{aligned}&\min_{\delta} (1+\delta)^{\frac{N}{K+1}} \left| \boldsymbol{S} + \frac{1}{(1+\delta)} (\boldsymbol{x}_0 - \hat{\alpha}\boldsymbol{s})(\boldsymbol{x}_0 - \hat{\alpha}\boldsymbol{s})^H \right| \\ &= \min_{\delta} |\boldsymbol{S}| (1+\delta)^{\frac{N}{K+1}-1} \left[1 + \delta + \tilde{\boldsymbol{x}}_0^H \tilde{\boldsymbol{x}}_0 - \tilde{\boldsymbol{x}}_0^H \tilde{\boldsymbol{s}} (\tilde{\boldsymbol{s}}^H \tilde{\boldsymbol{s}})^{-1} \tilde{\boldsymbol{s}}^H \tilde{\boldsymbol{x}}_0 \right]\end{aligned} \tag{4.14}$$

$$\hat{\delta} = \begin{cases} (\zeta - 1)T - 1, (\zeta - 1)T - 1 > 0 \\ 0, else \end{cases} \tag{4.15}$$

where $T = \tilde{\boldsymbol{x}}_0^H \boldsymbol{P}_{\tilde{s}}^{\perp} \tilde{\boldsymbol{x}}_0$, $\tilde{\boldsymbol{x}}_0 = \boldsymbol{S}^{-\frac{1}{2}} \boldsymbol{x}_0$, $\tilde{\boldsymbol{s}} = \boldsymbol{S}^{-\frac{1}{2}} \boldsymbol{s}$, $\boldsymbol{P}_{\tilde{s}}^{\perp} = \boldsymbol{I}_N - \tilde{\boldsymbol{s}} (\tilde{\boldsymbol{s}}^H \tilde{\boldsymbol{s}})^{-1} \tilde{\boldsymbol{s}}^H$, $\zeta = \frac{K+1}{N}$.

The adaptive complex parameter gradient detector can be obtained by plugging Equations 4.6–4.15 into Equation 4.4 and simplifying the result:

$$T_{Gradient} = 2 \,\mathrm{Re} \left(\frac{\boldsymbol{x}_0^H \hat{\boldsymbol{R}}_0^{-1} \boldsymbol{s}\boldsymbol{s}^H \boldsymbol{S}^{-1} \boldsymbol{x}_0}{\boldsymbol{s}^H \boldsymbol{S}^{-1} \boldsymbol{s}} \right) + \hat{\delta} \left[\mathrm{tr} \left(\boldsymbol{x}_0^H \hat{\boldsymbol{R}}_0^{-1} \boldsymbol{x}_0 \right) - N \right] \underset{H_0}{\overset{H_1}{\gtrless}} \eta \tag{4.16}$$

The proposed detector has the CFAR property with respect to the noise covariance matrix and the proof is shown in Appendix 4.A.

4.3 NUMERICAL EXAMPLES

Numerical examples are given to assess the detection performance of the new robust detector. We set $\boldsymbol{s} = \left[1, e^{j2\pi f_d}, \ldots, e^{j2\pi(N-1) f_d} \right]$, $j = \sqrt{-1}$, $f_d = 0.08$, $\boldsymbol{R} = \boldsymbol{M} + \sigma^2 \boldsymbol{I}_N$, $\boldsymbol{M}(i,j) = \tau^{(i-j)^2}$ denotes the Gaussian-shaped noise covariance matrix ,$\sigma^2 \boldsymbol{I}_N$ denotes a white noise covariance matrix 10 dB weaker than the Gaussian-shaped noise, $N=10$, $P_{fa}=10^{-3}$, $\tau=\exp\left(-2\pi^2\sigma_f^2\right)$, $\sigma_f=0.05$. The SNR is:

$$\mathrm{SNR}=|\alpha|^2 \boldsymbol{s}^H \boldsymbol{R}^{-1}\boldsymbol{s} \tag{4.17}$$

4.3.1 Mismatched Detection Performance Analysis

The mismatched detection performance of the new detector is analyzed. The detection performance of the conventional AMF [10], Kelly's GLRT [11], the parametric GLRT (PGLRT) detector proposed in [8], and the Rao and Wald detectors proposed in [9] is also given:

$$T_{AMF}=\frac{\left|\boldsymbol{s}^H\boldsymbol{S}^{-1}\boldsymbol{x}_0\right|^2}{\boldsymbol{s}^H\boldsymbol{S}^{-1}\boldsymbol{s}} \tag{4.18}$$

$$T_{GLRT}=\frac{\left|\boldsymbol{s}^H\boldsymbol{S}^{-1}\boldsymbol{x}_0\right|^2}{\left(1+\boldsymbol{x}_0^H\boldsymbol{S}^{-1}\boldsymbol{x}_0\right)\left(\boldsymbol{s}^H\boldsymbol{S}^{-1}\boldsymbol{s}\right)} \tag{4.19}$$

$$T_{PGLRT}=\begin{cases}\dfrac{\left(1+\left\|\tilde{\boldsymbol{x}}_0\right\|^2\right)\left(1-\dfrac{1}{\zeta_\varepsilon}\right)}{\left[\left(\zeta_\varepsilon-1\right)\left\|\boldsymbol{P}_{\tilde{s}}^{\perp}\tilde{\boldsymbol{x}}_0\right\|^2\right]^{\frac{1}{\zeta_\varepsilon}}}, & \left\|\boldsymbol{P}_{\tilde{s}}^{\perp}\tilde{\boldsymbol{x}}_0\right\|^2>\dfrac{1}{\zeta_\varepsilon-1}\\[2ex] \dfrac{1+\left\|\tilde{\boldsymbol{x}}_0\right\|^2}{1+\left\|\boldsymbol{P}_{\tilde{s}}^{\perp}\tilde{\boldsymbol{x}}_0\right\|^2}, & \text{otherwise}\end{cases} \tag{4.20}$$

$$T_{Wald}=\frac{2(K+1)}{1+\hat{\delta}}T_{AMF}+\frac{NK\hat{\delta}^2}{(K+1)\left(1+\hat{\delta}\right)^2} \tag{4.21}$$

$$T_{Rao}=2KT_{AMF}+\frac{(K+1)}{NK}\left[K\mathrm{tr}\left(\boldsymbol{S}^{-1}\boldsymbol{x}_0\boldsymbol{x}_0^H\right)-N\right]^2 \tag{4.22}$$

where $\zeta_\varepsilon=\dfrac{K+1}{N}(1+\varepsilon),\varepsilon\geq 0$.

The mismatch angle and the SNR are

$$\cos^2\theta = \left|\mathbf{s}^H \mathbf{R}^{-1}\mathbf{s}_a\right|^2 \Big/ \left[\left(\mathbf{s}^H \mathbf{R}^{-1}\mathbf{s}\right)\left(\mathbf{s}_a^H \mathbf{R}^{-1}\mathbf{s}_a\right)\right] \tag{4.23}$$

$$\text{SNR} = \left|\alpha\right|^2 \mathbf{s}_a^H \mathbf{R}^{-1}\mathbf{s}_a \tag{4.24}$$

where $\mathbf{s}_a = \left[1, e^{j2\pi f_{df}}, \ldots, e^{j2\pi(N-1)f_{df}}\right]$ denotes the actual signal steering vector, $f_{df} = f_d + \Delta/N$ denotes the Doppler frequency of the actual signal.

In Figure 4.1, the PD as a function of $\cos^2\theta$ is plotted to analyze the gradient detector's mismatched detection performance when $\text{SNR} = 20$ dB and $K = 30, 40$. Compared with the PGLRT, Wald, AMF, and Kelly's GLRT, the proposed gradient detector has higher probabilities of detection for small $\cos^2\theta$. That is to say, the gradient detector is more robust than the PGLRT, Wald, AMF, and Kelly's GLRT. Compared with the Rao detector, the proposed gradient detector achieves better robustness when $\cos^2\theta > 0.48$.

4.3.2 Matched Detection Performance Analysis

The matched detection performance of the adaptive gradient detector is tested in Figure 4.2. Compared with the Rao test, the gradient detector shows 2 dB matched detection performance gain. Compared with the PGLRT, AMF, Wald, and GLRT, the proposed gradient detector suffers about 1 dB matched detection performance loss. In conclusion, the gradient detector can achieve comparable robustness and improved matched detection performance compared to Rao test. The gradient detector guarantees stronger robustness than PGLRT, AMF, Wald, and GLRT in signal mismatch cases at the expense of a slight matched detection performance loss.

4.4 CONCLUSION

In this chapter, we considered the adaptive signal detection problem in Gaussian noise. The complex parameter gradient detector was proposed by introducing a random perturbation under the alternative hypothesis. The proposed gradient detector was proved to have a CFAR property with respect to the noise covariance matrix. Meanwhile, the detection performance analysis highlighted that the gradient detector can achieve better

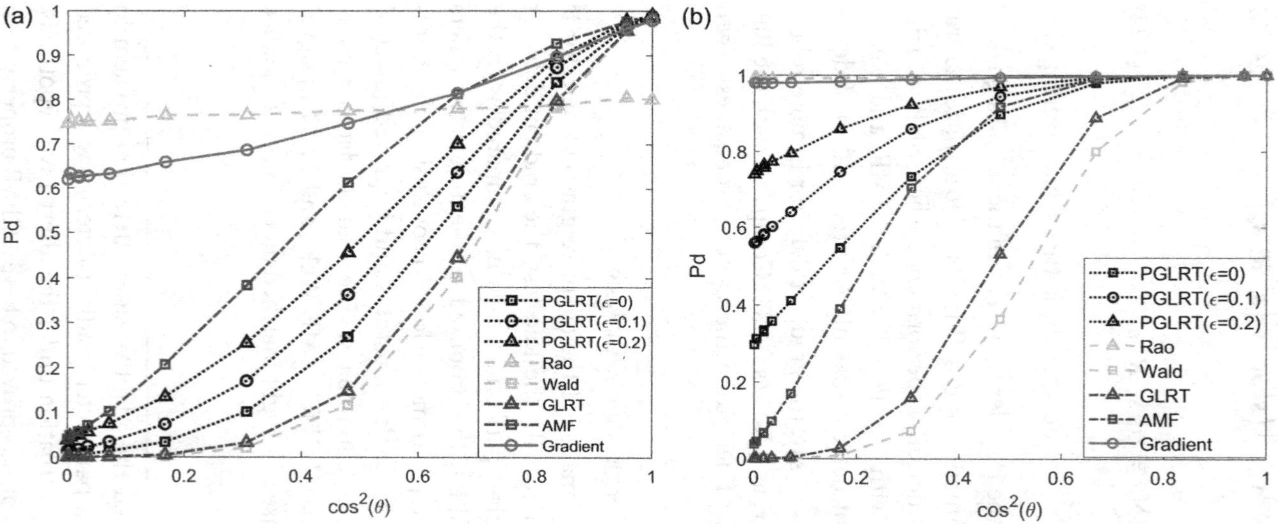

FIGURE 4.1 P_d versus $\cos^2\theta$ of the detectors for $N = 20$, SNR = 20dB: (a) $K = 30$; (b) $K = 40$.

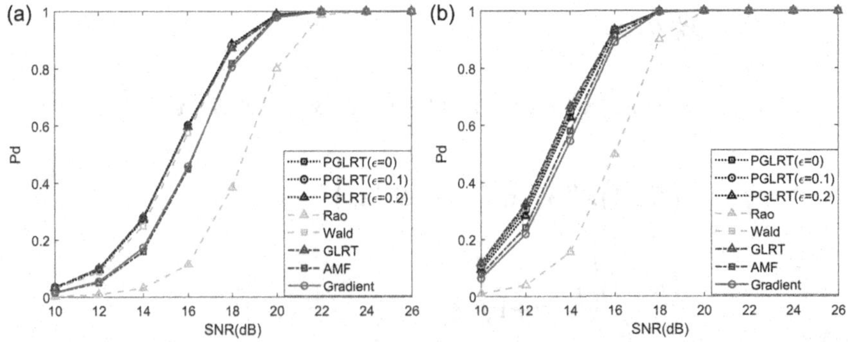

FIGURE 4.2 P_d versus SNR for $N = 20$, $\cos^2\theta = 1$: (a) $K = 30$; (b) $K = 40$.

robustness than its GLRT-based counterparts. A possible future research may be the investigation of designing robust detectors in non-Gaussian noise.

APPENDIX 4.A THE CFAR PROPERTY ANALYSIS

In this part, we prove the CFAR property of the gradient detector. According to the properties of the inverse matrix [12], we obtain

$$\begin{aligned}\hat{\boldsymbol{R}}_0^{-1} &= \left[\left(\boldsymbol{S} + \boldsymbol{x}_0\boldsymbol{x}_0^H\right)/(K+1)\right]^{-1} \\ &= (K+1)\left[\boldsymbol{S}^{-1} - \boldsymbol{S}^{-1}\boldsymbol{x}_0\left(1 + \boldsymbol{x}_0^H\boldsymbol{S}^{-1}\boldsymbol{x}_0\right)^{-1}\boldsymbol{x}_0^H\boldsymbol{S}^{-1}\right]\end{aligned} \tag{4.25}$$

Then, we have the following results:

$$\begin{aligned}&\frac{\boldsymbol{x}_0^H\hat{\boldsymbol{R}}_0^{-1}\boldsymbol{s}\boldsymbol{s}^H\boldsymbol{S}^{-1}\boldsymbol{x}_0}{\boldsymbol{s}^H\boldsymbol{S}^{-1}\boldsymbol{s}} \\ &= (K+1)\left[1 - \boldsymbol{x}_0^H\boldsymbol{S}^{-1}\boldsymbol{x}_0\left(1 + \boldsymbol{x}_0^H\boldsymbol{S}^{-1}\boldsymbol{x}_0\right)^{-1}\right]\boldsymbol{x}_0^H\boldsymbol{S}^{-1}\boldsymbol{s}\boldsymbol{s}^H\boldsymbol{S}^{-1}\boldsymbol{x}_0\left(\boldsymbol{s}^H\boldsymbol{S}^{-1}\boldsymbol{s}\right)^{-1} \\ &= (K+1)\frac{\boldsymbol{x}_0^H\boldsymbol{S}^{-1}\boldsymbol{s}\boldsymbol{s}^H\boldsymbol{S}^{-1}\boldsymbol{x}_0}{\left(1 + \boldsymbol{x}_0^H\boldsymbol{S}^{-1}\boldsymbol{x}_0\right)\left(\boldsymbol{s}^H\boldsymbol{S}^{-1}\boldsymbol{s}\right)} \\ &= (K+1)T_{GLRT} = (K+1)\frac{\gamma}{1+\gamma}\end{aligned} \tag{4.26}$$

$$\begin{aligned}
&\boldsymbol{x}_0^H \hat{\boldsymbol{R}}_0^{-1} \boldsymbol{x}_0 \\
&= (K+1)\boldsymbol{x}_0^H \boldsymbol{S}^{-1}\boldsymbol{x}_0 \left[1-\left(1+\boldsymbol{x}_0^H \boldsymbol{S}^{-1}\boldsymbol{x}_0\right)^{-1} \boldsymbol{x}_0^H \boldsymbol{S}^{-1}\boldsymbol{x}_0\right] \\
&= (K+1)\boldsymbol{x}_0^H \boldsymbol{S}^{-1}\boldsymbol{x}_0 \left(1+\boldsymbol{x}_0^H \boldsymbol{S}^{-1}\boldsymbol{x}_0\right)^{-1} \\
&= (K+1)\left(1-\frac{\varsigma}{\gamma+1}\right)
\end{aligned} \tag{4.27}$$

$$\begin{aligned}
T &= \tilde{\boldsymbol{x}}_0^H \boldsymbol{P}_{\tilde{s}}^{\perp} \tilde{\boldsymbol{x}}_0 \\
&= \boldsymbol{x}_0^H \boldsymbol{S}^{-1}\boldsymbol{x}_0 - \boldsymbol{x}_0^H \boldsymbol{S}^{-1}\boldsymbol{s}\left(\boldsymbol{s}^H \boldsymbol{S}^{-1}\boldsymbol{s}\right)^{-1} \boldsymbol{s}^H \boldsymbol{S}^{-1}\boldsymbol{x}_0 \\
&= \frac{1}{\varsigma}-1
\end{aligned} \tag{4.28}$$

where $\gamma = \dfrac{\boldsymbol{x}_0^H \boldsymbol{S}^{-1}\boldsymbol{s}\boldsymbol{s}^H \boldsymbol{S}^{-1}\boldsymbol{x}_0}{\boldsymbol{s}^H \boldsymbol{S}^{-1}\boldsymbol{s}} \Big/ \left(1+\boldsymbol{x}_0^H \boldsymbol{S}^{-1}\boldsymbol{x}_0 - \dfrac{\boldsymbol{x}_0^H \boldsymbol{S}^{-1}\boldsymbol{s}\boldsymbol{s}^H \boldsymbol{S}^{-1}\boldsymbol{x}_0}{\boldsymbol{s}^H \boldsymbol{S}^{-1}\boldsymbol{s}}\right)$,

$\varsigma = 1 \Big/ \left(1+\boldsymbol{x}_0^H \boldsymbol{S}^{-1}\boldsymbol{x}_0 - \dfrac{\boldsymbol{x}_0^H \boldsymbol{S}^{-1}\boldsymbol{s}\boldsymbol{s}^H \boldsymbol{S}^{-1}\boldsymbol{x}_0}{\boldsymbol{s}^H \boldsymbol{S}^{-1}\boldsymbol{s}}\right)$. It is found that $\varsigma \sim \mathcal{CB}_{K-N+2,N-1}$ and $\gamma \sim \mathcal{CF}_{1,K-N+1}$ [8] are all independent with noise covariance matrix, where $\mathcal{CB}$ and $\mathcal{CF}$ denote complex central beta distribution and complex central F-distribution, respectively. Since the test statistic of the gradient detector Equation 4.16 is the function of $\dfrac{\boldsymbol{x}_0^H \hat{\boldsymbol{R}}_0^{-1}\boldsymbol{s}\boldsymbol{s}^H \boldsymbol{S}^{-1}\boldsymbol{x}_0}{\boldsymbol{s}^H \boldsymbol{S}^{-1}\boldsymbol{s}}$, $\boldsymbol{x}_0^H \hat{\boldsymbol{R}}_0^{-1}\boldsymbol{x}_0$, and $T = \tilde{\boldsymbol{x}}_0^H \boldsymbol{P}_{\tilde{s}}^{\perp} \tilde{\boldsymbol{x}}_0$, the proposed detector is CFAR with respect to the noise covariance matrix.

REFERENCES

1. O. Besson, "Adaptive Detection with Bounded Steering Vectors Mismatch Angle," *IEEE Transactions on Signal Processing*, vol. 55, pp. 1560–1564, 2007.
2. S. Lee, M. Nguyen, I. Song, J. Bae, and S. Yoon, "Detection Schemes for Range-Spread Targets Based on The Semidefinite Problem," *IEEE Transactions on Aerospace and Electronic Systems*, vol. 55, no. 1, pp. 57–69, 2019.

3. R. S. Raghavan, N. Pulsone, and D. J. McLaughlin, "Performance of the GLRT for Adaptive Vector Subspace Detection," *IEEE Transactions on Aerospace and Electronic Systems*, vol. 32, no. 4, pp. 1473–1487, 1996.
4. J. Liu, S. Sun, and W. Liu, "One-Step Persymmetric GLRT for Subspace Signals," *IEEE Transactions on Signal Processing*, vol. 67, no. 14, pp. 3639–3648, July 15, 2019.
5. L. Cai and H. Wang, "A Persymmetric Multiband GLR Algorithm," *IEEE Transactions on Aerospace and Electronic Systems*, vol. 28, no. 3, pp. 806–816, 1992.
6. A. De Maio, "Robust Adaptive Radar Detection in the Presence of Steering Vector Mismatches," *IEEE Transactions on Aerospace and Electronic Systems*, vol. 41, no. 4, pp. 1322–1337, 2005.
7. E. Conte, A. De Maio, and G. Ricci, "GLRT-based Adaptive Detection Algorithm for Range-Spread Targets," *IEEE Transactions on Signal Processing*, vol. 49, no. 7, pp. 1336–1348, 2001.
8. A. Coluccia, G. Ricci, and O. Besson, "Design of Robust Radar Detectors Through Random Perturbation of the Target Signature," *IEEE Transactions on Signal Processing*, vol. 67, no. 19, pp. 5118–5129, 2019.
9. S. Sun, J. Liu, W. Liu, and T. Jian, "Robust Detection of Distributed Targets Based on Rao Test and Wald Test," *Signal Processing*, vol. 180, pp. 1–10, 2021.
10. F. C. Robey, D. R. Fuhrmann, E. J. Kelly, and R. Nitzberg, "A CFAR Adaptive Matched Filter Detector," *IEEE Transactions on Aerospace and Electronic Systems*, vol. 28, no. 1, pp. 208–216, 1992.
11. E. J. Kelly, "An Adaptive Detection Algorithm," *IEEE Transactions on Aerospace and Electronic Systems*, vol. AES-22, no. 2, pp. 115–127, 1986.
12. E. J. Kelly and K. Forsythe, *Adaptive Detection and Parameter Estimation for Multidimensional Signal Models*, Lexington: Lincoln Laboratory, Tech. Rep. 848, 1989.

CHAPTER 5

Tunable Adaptive Detector for Mismatched Signals

5.1 INTRODUCTION

As discussed in Chapters 3 and 4, in practical applications, the actual signal backscattered from a target can be different from the nominal one. This phenomenon is often called signal mismatch, which can be caused due to imperfect array calibration, spatial multipath, pointing errors, and so on [1, 2]. In order to improve the mismatched signals rejection, the ABORT [3], and whitened ABORT (W-ABORT) [4] are proposed. The null hypothesis is modified by adding a fictitious signal that is orthogonal to the signal steering vector in the whitened or quasi-whitened space. If a target exists in a direction different from the nominal one, the detector will incline to the null hypothesis. In [5, 6], a different decision scheme is introduced. Precisely, at the design stage, it is assumed that the CUT contains a noise-like interferer, which is not accounted for in the training data. Consequently, the double-normalized AMF (DN-AMF) is proposed therein. In [7–9], two-stage detectors, i.e., the adaptive sidelobe blanker (ASB) and its improved versions, are proposed. These detectors usually contain two individual detectors with converse capabilities of mismatched signals rejection. The detectors declare the presence of a target in the CUT only when data survive both detection thresholds. Another scheme is modelling the actual signal as a vector belonging to a proper

DOI: 10.1201/9781003477907-5

cone [10], and then devising the detector by the theory of convex optimization. The other effective approach is resorting to tunable detectors [1, 2, 11–16]. By tuning a scalar parameter, they can control the level to which the mismatched signals are rejected.

The ABORT, W-ABORT, and DN-AMF all have enhanced mismatched signals rejection, but lack robustness to slightly mismatched signals. The capabilities of rejection and robustness to mismatched signals of the ASB-like detectors are confined by these capabilities of their individual detectors. As for the techniques of modelling the actual signal as a vector belonging to a proper cone, it only possesses robustness to the mismatched signals, and it usually has no closed-form solution. Moreover, the tunable detector in [11] has limited capabilities of mismatched signals rejection, while the tunable detectors in [12, 13] has limited robustness.

In this chapter, by comparing the similarity of the KGLRT, AMF, and ACE, we introduce a novel parametric detector, which encompasses the three aforementioned detectors as its special cases. For this reason, the novel detector is denoted as the KMACE. The first and second letters in the acronym above stand for the KGLRT and AMF, respectively, while the last three letters are the ACE. Unlike the single-parameter tunable detectors in [1, 2, 11–16], the KMACE is parameterized by two tunable parameters. Precisely, one parameter is referred to as the additive parameter and the other is denoted as the exponential parameter. Remarkably, the KMACE has superior flexibility in controlling the degree to which the mismatched signals are rejected. Increasing the additive parameter and/or decreasing the exponential parameter makes the KMACE much more robust to the signal mismatch. In contrast, decreasing the additive parameter and/or increasing the exponential parameter makes the KMACE much more selective (less tolerant to the signal mismatch). Moreover, the KMACE can also provide improved detection performance for matched signals than the KGLRT, AMF, and ACE.

The rest of this chapter is organized in the following fashion. Section 5.2 briefly describes the adaptive detection problem and presents the new parametric detector, while Section 5.3 shows the performance assessment. The numerical examples are given in Section 5.4. Finally, Section 5.5 concludes the chapter.

5.2 PROBLEM FORMULATION AND DETECTOR DESIGN

We denote the primary data by an N-dimensional column vector $\boldsymbol{x}$. We want to discriminate between hypothesis H_0 that $\boldsymbol{x}$ only contains the disturbance and hypothesis H_1 that $\boldsymbol{x}$ contains the disturbance and a useful signal. Hence, the detection problem can be formulated as the following binary hypothesis test:

$$\begin{cases} H_0 : \boldsymbol{x} = \boldsymbol{n} \\ H_1 : \boldsymbol{x} = a\boldsymbol{s} + \boldsymbol{n} \end{cases} \tag{5.1}$$

where a is the signal amplitude, $\boldsymbol{s}$ is the (spatial, temporal, or spatial-temporal) signal steering vector, and $\boldsymbol{n}$ is the disturbance, containing clutter and white noise, with a positive definite covariance matrix $\boldsymbol{R}$. Note that the value of a and $\boldsymbol{R}$ are both unknown. As customary, we assume that a set of training data, $\boldsymbol{x}_l$, $l = 1, 2, \ldots, L$, is available, which is statistically independent and shares the same statistical property with the noise $\boldsymbol{n}$ in the primary data $\boldsymbol{x}$.

The AMF [17], KGLRT [18], and ACE [19] for the detection problem in (5.1) are

$$t_{\text{AMF}} = \frac{\left|\boldsymbol{s}^H \boldsymbol{S}^{-1} \boldsymbol{x}\right|^2}{\boldsymbol{s}^H \boldsymbol{S}^{-1} \boldsymbol{s}} \tag{5.2}$$

$$t_{\text{KGLRT}} = \frac{t_{\text{AMF}}}{1 + \boldsymbol{x}^H \boldsymbol{S}^{-1} \boldsymbol{x} - t_{\text{AMF}}} \tag{5.3}$$

and

$$t_{\text{ACE}} = \frac{t_{\text{AMF}}}{\boldsymbol{x}^H \boldsymbol{S}^{-1} \boldsymbol{x}} \tag{5.4}$$

respectively, where $\boldsymbol{S} = \boldsymbol{X}\boldsymbol{X}^H$ is L times the sample covariance matrix with $\boldsymbol{X} = [\boldsymbol{x}_1, \boldsymbol{x}_2, \ldots, \boldsymbol{x}_L]$ being the training data matrix.

By comparing (5.2)–(5.4), we introduce the KMACE as

$$t_{\text{KMACE}} = \frac{t_{\text{AMF}}}{\alpha + (\boldsymbol{x}^H \boldsymbol{S}^{-1} \boldsymbol{x} - t_{\text{AMF}})^{\gamma}} \tag{5.5}$$

where the additive parameter α and the exponential parameter γ are two positive scalars, referred to as the tunable parameters.

When $\alpha = \gamma = 0$, (5.5) reduces to the AMF. When $\alpha = \gamma = 1$, (5.5) degenerates into the KGLRT. Furthermore, when $\alpha = 0$ and $\gamma = 1$, (5.5) becomes

$$t_{\text{KMACE}} = \frac{t_{\text{AMF}}}{\boldsymbol{x}^H \boldsymbol{S}^{-1} \boldsymbol{x} - t_{\text{AMF}}} = \frac{1}{t_{\text{ACE}}^{-1} - 1} \tag{5.6}$$

where t_{ACE} is given in (5.4). Note that Equation 5.6 can be taken as a monotonically increasing function of t_{ACE} in the interval $(0,1)$. Moreover, $t_{\text{ACE}} \in (0,1)$. Hence, (5.6) is statistically equivalent[1] to t_{ACE}. In summary, when $\alpha = 0$ and $\gamma = 1$, Equation 5.5 turns into the ACE.

Note that t_{AMF} can be expressed as $\tilde{\boldsymbol{x}}^H \boldsymbol{P}_{\tilde{s}} \tilde{\boldsymbol{x}}$, while the quantity $\boldsymbol{x}^H \boldsymbol{S}^{-1} \boldsymbol{x} - t_{\text{AMF}}$ can be recast as $\tilde{\boldsymbol{x}}^H \boldsymbol{P}_{\tilde{s}}^{\perp} \tilde{\boldsymbol{x}}$, where $\tilde{\boldsymbol{x}} = \boldsymbol{S}^{-1/2} \boldsymbol{x}$, $\tilde{\boldsymbol{s}} = \boldsymbol{S}^{-1/2} \boldsymbol{s}$, $\boldsymbol{S}^{1/2}$ is the square-root matrix of $\boldsymbol{S}$, $\boldsymbol{S}^{-1/2}$ is the inversion of $\boldsymbol{S}^{1/2}$, $\boldsymbol{P}_{\tilde{s}}$ is the orthogonal projection matrix (projector) onto the column space of $\tilde{\boldsymbol{s}}$, and $\boldsymbol{P}_{\tilde{s}}^{\perp}$ is the orthogonal complement of $\boldsymbol{P}_{\tilde{s}}$, i.e., $\boldsymbol{P}_{\tilde{s}}^{\perp} = \boldsymbol{I} - \boldsymbol{P}_{\tilde{s}}$. It follows that t_{AMF} can be taken as the energy of the quasi-whitened primary data $\tilde{\boldsymbol{x}}$ projected onto the quasi-whitened signal subspace spanned by $\tilde{\boldsymbol{s}}$, denoted as $<\tilde{\boldsymbol{s}}>$, while $\boldsymbol{x}^H \boldsymbol{S}^{-1} \boldsymbol{x} - t_{\text{AMF}}$ is the energy of $\tilde{\boldsymbol{x}}$ projected onto the complementary subspace of $<\tilde{\boldsymbol{s}}>$, i.e., $<\tilde{\boldsymbol{s}}>^{\perp}$. Hence, we can reasonably conjecture that increasing the value of α and fixing the value of γ, will weaken the effect of the term $\boldsymbol{x}^H \boldsymbol{S}^{-1} \boldsymbol{x} - t_{\text{AMF}}$. Consequently, the detector turns more and more robust to mismatched signals. On the other hand, by increasing the value of γ and fixing the value of α, the capabilities of the rejection of mismatched signals will increase. This is indeed the case, as shown in Section 5.4.

5.3 PERFORMANCE EVALUATION

We now proceed to analytically assess the performance of the KMACE in terms of PFA and PD. The PD of the KMACE can be expressed as

$$\Pr(t_{\text{KMACE}} > \eta; \text{H}_1) \tag{5.7}$$

where η is the threshold to be assigned to ensure a presumed PFA.

We consider the case of signal mismatch. When this phenomenon arises, the actual signal becomes $a\boldsymbol{s}_0$, where $\boldsymbol{s}_0$ is the actual signal steering vector, which is not necessarily aligned with the nominal array steering vector $\boldsymbol{s}$. To quantify the signal mismatch, we introduce the quantity [3]

$$\cos^2\phi = \frac{|\boldsymbol{s}^H \boldsymbol{R}^{-1} \boldsymbol{s}_0|^2}{\boldsymbol{s}^H \boldsymbol{R}^{-1} \boldsymbol{s} \cdot \boldsymbol{s}_0^H \boldsymbol{R}^{-1} \boldsymbol{s}_0} \tag{5.8}$$

which is the cosine squared of the array steering $\boldsymbol{s}$ and the actual signal steering $\boldsymbol{s}_0$ in the whitened space.

It is straightforward to verify that Equation 5.5 can be recast as

$$t_{\text{KMACE}} = \frac{t_{\text{KGLRT}}\beta^{\gamma-1}}{\alpha\beta^{\gamma} + (1-\beta)^{\gamma}} \tag{5.9}$$

where

$$\beta = (1 + \boldsymbol{x}^H \boldsymbol{S}^{-1}\boldsymbol{x} - t_{\text{AMF}})^{-1} \tag{5.10}$$

and we have used the fact $t_{\text{KGLRT}} = t_{\text{AMF}}\beta$, with t_{KGLRT} being the KGLRT given in (5.3).

The conditional statistical distribution of t_{KGLRT} with β fixed under H_1, is $t_{\text{KGLRT}}\,|\,\beta, H_1 \sim \mathcal{CF}_{1,L-N+1}(\rho_{\phi}\beta)\ \rho_{\phi}\beta$ [22], where $\rho_{\phi} = \rho\cos^2\phi$ and

$$\rho = |a|^2\, \boldsymbol{s}_0^H \boldsymbol{R}^{-1}\boldsymbol{s}_0 \tag{5.11}$$

is the output signal-to-clutter-plus-noise ratio (SCNR). Hence, plugging (5.9) into (5.7) results in

$$\begin{aligned} \text{PD} &= \Pr\left[t_{\text{KGLRT}} > \eta\beta^{1-\gamma}[\alpha\beta^{\gamma} + (1-\beta)^{\gamma}]; \text{H}_1\right] \\ &= \int_0^1 \left(1 - \mathcal{P}_{1|\beta}\left\{\eta\beta^{1-\gamma}[\alpha\beta^{\gamma} + (1-\beta)^{\gamma}]\right\}\right) f_1(\beta)d\beta \end{aligned} \tag{5.12}$$

where $\mathcal{P}_{1|\beta}(\cdot)$ is the cumulative distribution function (CDF) of t_{KGLRT} for given β under H_1, i.e., [22]

$$\mathcal{P}_{1|\beta}(\eta) = \frac{\eta}{(1+\eta)^{L-N+1}} \sum_{k=0}^{L-N} \binom{L-N+1}{1+k} \eta^k \cdot \text{IG}_{k+1}\left(\frac{\rho_{\phi}\beta}{1+\eta}\right) \tag{5.13}$$

where $\text{IG}_{m+1}(a)$ is the incomplete Gamma function, given by $\text{IG}_{m+1}(a) = e^{-a}\sum_{k=0}^{m} a^k/k!$. Moreover, β is distributed as $\beta \sim \mathcal{CB}_{L-N+2,N-1}(\delta^2)$, $\delta^2 = \rho\sin^2\phi$ [22]. The quantity $f_1(\beta)$ in (5.12) is the PDF of β under H_1, which is found to be [22]

$$f_1(\beta)=e^{-\delta^2\beta}\sum_{m=0}^{L-N+2}\binom{L-N+2}{m}\frac{L!\delta^{2m}}{(L+m)!}f_{L-N+2,N+m-1}(\beta) \tag{5.14}$$

where $f_{m,n}(\beta)$ is the PDF of the complex central Beta distribution with m and n degrees of freedom (DOFs).

Setting $\rho_\phi=0$ in (5.12), we have the PFA as

$$\text{PFA}=\int_0^1\left[1+\eta\alpha\beta+\eta\beta^{1-\gamma}(1-\beta)^\gamma\right]^{-(L-N+1)}f_0(\beta)d\beta \tag{5.15}$$

where we have used the identity $1-\mathcal{P}_{0|\beta}(x)=(1+x)^{-(L-N+1)}$ [12], $\mathcal{P}_{0|\beta}(\cdot)$ is the CDF of t_{KGLRT} for given β under H_0, and $f_0(\beta)$ is the PDF of β under H_0. $f_0(\beta)$ can be obtained by setting $\delta^2=0$ in (5.14).

From (5.15), we know the PFA is not dependent on the noise covariance matrix $\boldsymbol{R}$. Hence, the KMACE is CFAR.

5.4 NUMERICAL EXAMPLES

In this section, we evaluate the detection performance of the KMACE both for matched and mismatched signals. For independent confirmation, we also show the PD of the KMACE obtained by Monte Carlo simulations, for which 10^4 and $100/\text{PFA}$ independent data realizations are run to calculate the PD and the detection threshold necessary to ensure a preassigned PFA. The dimension of the primary data is set to be $N=12$, the number of the training data is chosen as $L=2N$, and the PFA is selected as $\text{PFA}=10^{-3}$.

Figure 5.1 depicts the PD of the KMACE with different tunable parameters for matched signals. The plane, drawn in Figure 5.1, is the PD of the KGLRT. For clear display, the PDs of the KMACE less than 0.9 are set to be 0.9. It is shown that it is about one-third of the number of couples of γ and α, with which the PD of the KMACE is higher than that of the KGLRT. Furthermore, for fixed γ with a small value, the value of α does not significantly affect the PD of the KMACE, while for fixed γ with a large value, with the increase of the value of α, the PD turns higher and higher. On the other hand, when the value of α is fixed, the increase of the value of γ results in lower PD, but it will bring enhanced rejection of mismatched signals; see Figures 5.3– 5.5.

Figure 5.2 plots the PD of the KMACE for matched signals under different SCNRs, also in comparison with those of the KGLRT, AMF, and ACE.

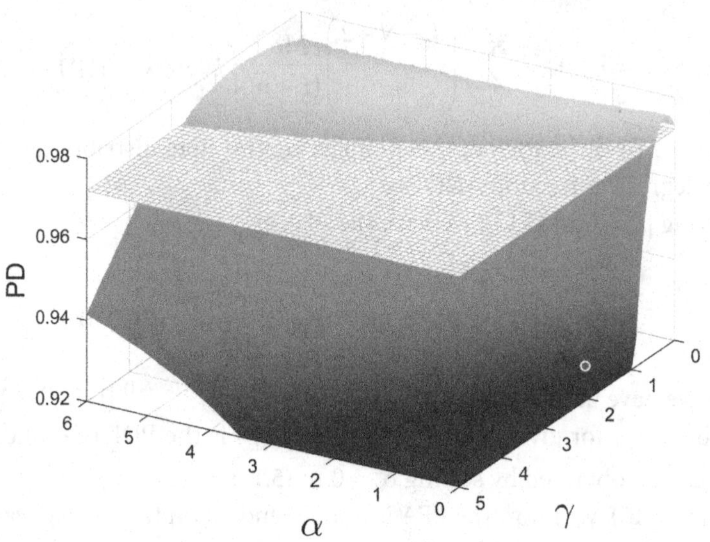

FIGURE 5.1 PDs of the KMACE with different tunable parameters for matched signals. SCNR=16 dB.

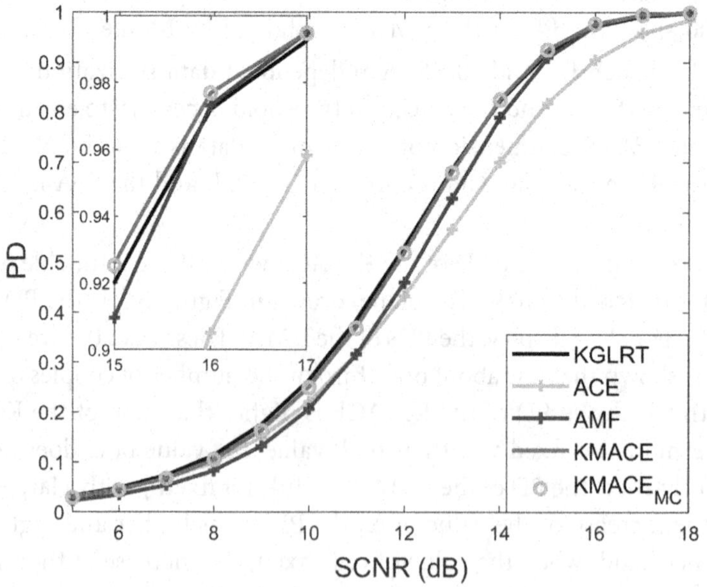

FIGURE 5.2 PDs of the detectors for matched signals.

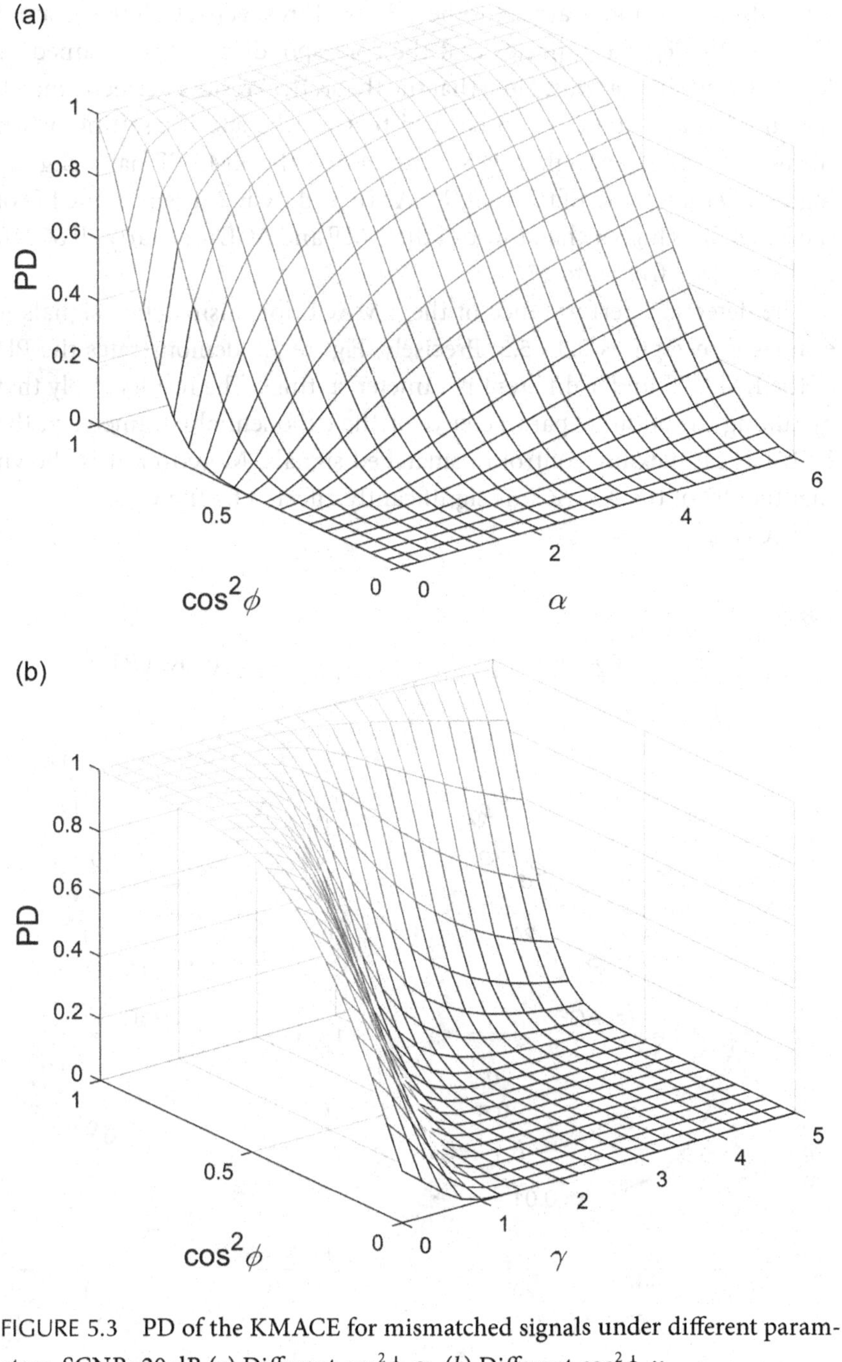

FIGURE 5.3 PD of the KMACE for mismatched signals under different parameters. SCNR=20 dB (*a*) Different $\cos^2\phi$-α. (*b*) Different $\cos^2\phi$-γ.

The values of α and γ are set to be 2.3 and 1, respectively. The subscript "MC" in the legend indicates that the corresponding PD is obtained by Monte Carlo simulations. Note that the theoretical results perfectly match the simulation results. It can be seen that for the specific setting, when the SCNR is high enough (say, SCNR>14 dB), the KMACE has a slightly higher PD than those of the KGLRT, AMF, and ACE. Moreover, the PD of the KMACE is higher than those of the AMF and ACE for nearly all SCNR in the interval from 7 to 17.

The detection performance of the KMACE for mismatched signals is evaluated in Figures 5.3– 5.5. Precisely, Figure 5.3 demonstrates the PD of the KMACE under different parameter settings. The results imply that by tuning the additive parameter α or the exponential parameter γ, the KMACE can flexibly control mismatched signals. Moreover, it is shown that the PD of the KMACE is significantly affected by the values of α, γ, and SCNR.

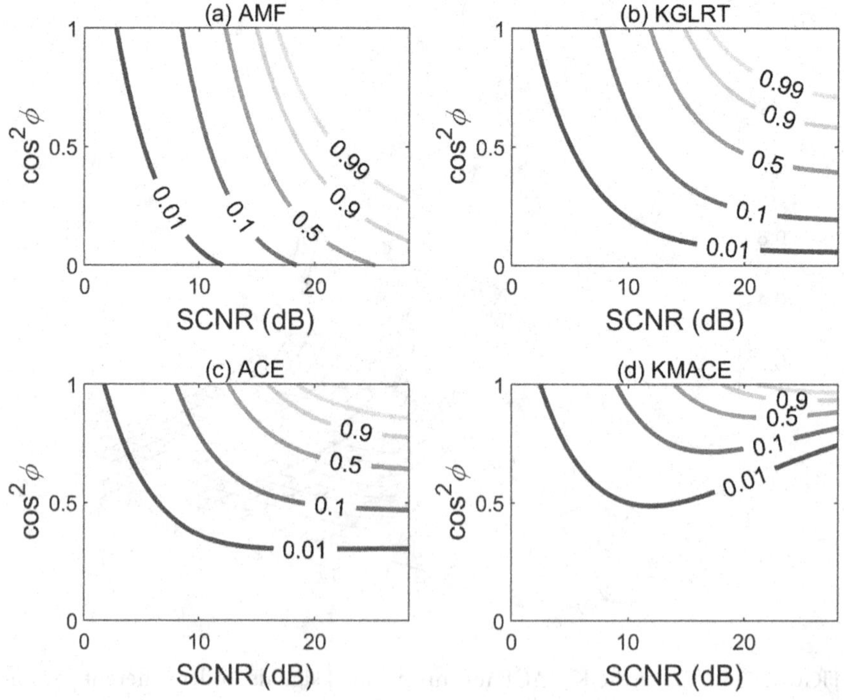

FIGURE 5.4 Contours of constant PDs for the AMF (subplot a), KGLRT (subplot b), ACE (subplot c), and KMACE (subplot d).

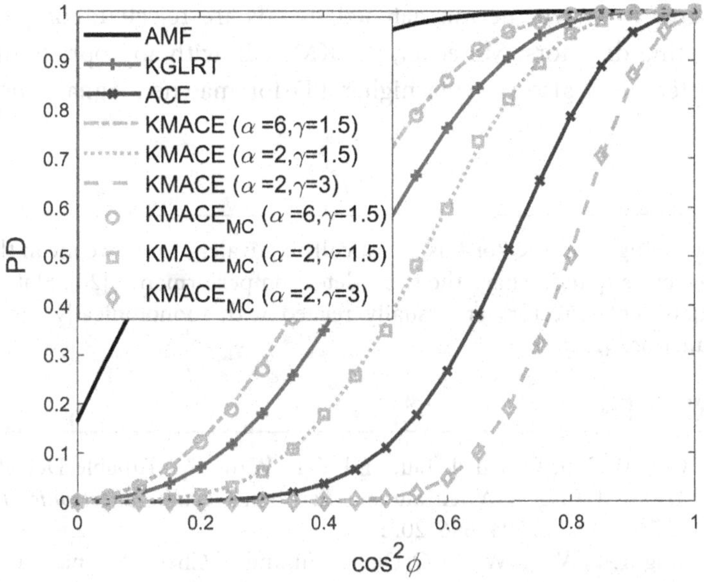

FIGURE 5.5 PD versus $\cos^2 \phi$ for mismatched signals. SCNR=20 dB.

Figure 5.4 plots contours of constant PDs, represented as functions of SCNR and $\cos^2 \phi$. Such plots are referred to as the mesh-plot [3]. For the KMACE, $\alpha = 0$ and $\gamma = 1.5$. It is seen that the KMACE has the best performance in terms of mismatched signals rejection, followed in sequence by the ACE, KGLRT, and AMF.

Figure 5.5 illustrates the PDs of the detectors versus $\cos^2 \phi$. The results indicate that by tuning the additive parameter α or/and the exponential parameter γ, the KMACE can achieve flexible control of the degree to which the mismatched signal is to be rejected. Precisely, fixing the value of γ to be 1.5, increasing the value of α from 2 to 6, the KMACE becomes more robust, while fixing the value of α to be 2, increasing the value of γ form 1.5 to 3, the KMACE turns more selective, and it is even more selective than the ACE. Besides, with $\alpha = \gamma = 0$, the KMACE degenerates to the AMF, and it has the least mismatch discrimination capabilities, or, equivalently, it is most robust to mismatched signals.

5.5 CONCLUSION

In this chapter, for the mismatched signal detection, we have proposed the tunable detector KMACE, which is doubly parameterized by additive and exponential parameters. The KMACE is more flexible in controlling the degree to which the mismatched signals are rejected, compared to the existing detectors. Moreover, the KMACE, with appropriate tunable parameters, can also provide higher PD for matched signals than the existing detectors.

NOTE

1. By saying "the detector t_1 is statistically equivalent to t_2", we mean that the detectors t_1 and t_2 have the same detection performance [20]. Statistically equivalent detectors are usually related with monotonically increasing functions [21].

REFERENCES

1. Y. Cui, W. Liu, Q. Du, J. Liu, and Y.-L. Wang, "A Tunable Detector for Distributed Targets When Signal Mismatch Occurs," *Electronics Letters*, vol. 57, no. 15, pp. 594–596, 2021.
2. P. Tang, Y.-L. Wang, W. Liu, Q. Du, C. Wu, and W. Chen, "A Tunable Detector for Distributed Target Detection in the Situation of Signal Mismatch," *IEEE Signal Processing Letters*, vol. 27, pp. 151–155, 2020.
3. N. B. Pulsone and C. M. Rader, "Adaptive Beamformer Orthogonal Rejection Test," *IEEE Transactions on Signal Processing*, vol. 49, no. 3, pp. 521–529, 2001.
4. F. Bandiera, O. Besson, and G. Ricci, "An ABORT-like Detector with Improved Mismatched Signals Rejection Capabilities," *IEEE Transactions on Signal Processing*, vol. 56, no. 1, pp. 14–25, 2008.
5. D. Orlando and G. Ricci, "A Rao Test with Enhanced Selectivity Properties in Homogeneous Scenarios," *IEEE Transactions on Signal Processing*, vol. 58, no. 10, pp. 5385–5390, 2010.
6. W. Liu, W. Xie, and Y.-L. Wang, "A Wald Test with Enhanced Selectivity Properties in Homogeneous Environments," *EURASIP Journal on Advances in Signal Processing*, vol. 2013, no. 14, pp. 1–4, 2013.
7. C. D. Richmond, "Performance of the Adaptive Sidelobe Blanker Detection Algorithm in Homogeneous Environments," *IEEE Transactions on Signal Processing*, vol. 48, no. 5, pp. 1235–1247, 2000.
8. F. Bandiera, O. Besson, D. Orlando, and G. Ricci, "An Improved Adaptive Sidelobe Blanker," *IEEE Transactions on Signal Processing*, vol. 56, no. 9, pp. 4152–4161, 2008.
9. K. Duan, M. Liu, H. Dai, F. Xu, and W. Liu, "A Two-Stage Detector for Mismatched Subspace Signals," *IEEE Geoscience and Remote Sensing Letters*, vol. 14, no. 12, pp. 2270–2274, 2017, doi: 10.1109/LGRS.2017.2761782.
10. A. De Maio, S. De Nicola, Y. Huang, S. Zhang, and A. Farina, "Adaptive Detection and Estimation in the Presence of Useful Signal and Interference Mismatches," *IEEE Transactions on Signal Processing*, vol. 57, no. 2, pp. 436–450, 2009.

11. S. Z. Kalson, "An Adaptive Array Detector with Mismatched Signal Rejection," *IEEE Transactions on Aerospace and Electronic Systems*, vol. 28, no. 1, pp. 195–207, 1992.
12. F. Bandiera, D. Orlando, and G. Ricci, "One- and Two-Stage Tunable Receivers*," *IEEE Transactions on Signal Processing*, vol. 57, no. 8, pp. 3264–3273, 2009.
13. C. Hao, B. Liu, S. Yan, and L. Cai, "Parametric Adaptive Radar Detector with Enhanced Mismatched Signals Rejection Capabilities," *EURASIP Journal on Advances in Signal Processing*, vol. 2010, pp. 1–11, 2010.
14. A. Zaimbashi and J. Li, "Tunable Adaptive Target Detection with Kernels in Colocated MIMO Radar," *IEEE Transactions on Signal Processing*, vol. 68, pp. 1500–1514, 2020, doi: 10.1109/TSP.2020.2975371.
15. J. Liu, S. Zhou, W. Liu, J. Zheng, H. Liu, and J. Li, "Tunable Adaptive Detection in Colocated MIMO Radar," *IEEE Transactions on Signal Processing*, vol. 66, no. 4, pp. 1080–1092, 2018, doi: 10.1109/TSP.2017.2778693.
16. A. Coluccia and G. Ricci, "A Tunable W-ABORT-like Detector with Improved Detection vs Rejection Capabilities," *IEEE Signal Processing Letters*, vol. 22, no. 6, pp. 713–717, 2015.
17. F. C. Robey, D. R. Fuhrmann, E. J. Kelly, and R. Nitzberg, "A CFAR Adaptive Matched Filter Detector," *IEEE Transactions on Aerospace and Electronic Systems*, vol. 28, no. 1, pp. 208–216, 1992.
18. E. J. Kelly, "An Adaptive Detection Algorithm," *IEEE Transactions on Aerospace and Electronic Systems*, vol. 22, no. 1, pp. 115–127, 1986.
19. S. Kraut and L. L. Scharf, "The CFAR Adaptive Subspace Detector is a Scale-Invariant GLRT," *IEEE Transactions on Signal Processing*, vol. 47, no. 9, pp. 2538–2541, 1999.
20. A. De Maio, S. M. Kay, and A. Farina, "On the Invariance, Coincidence, and Statistical Equivalence of the GLRT, Rao Test, and Wald Test," *IEEE Transactions on Signal Processing*, vol. 58, no. 4, pp. 1967–1979, 2010.
21. W. Liu, J. Liu, C. Hao, Y. Gao, and Y.-L. Wang, "Multichannel Adaptive Signal Detection: Basic Theory and Literature Review," *Science China: Information Sciences*, vol. 65, no. 2, pp. 121301, 2022, doi: 10.1007/s11432-020-3211-8.
22. E. J. Kelly, "Performance of an Adaptive Detection Algorithm: Rejection of Unwanted Signals," *IEEE Transactions on Aerospace and Electronic Systems*, vol. 25, no. 2, pp. 122–133, 1989.

CHAPTER 6

Adaptive Detection of a Subspace Signal in Interference

In a general sense, including clutter, thermal noise, and the possible target signal, the received data always include interference caused by unintentional industrial production like the radio and television signals, intentional electronic countermeasures, and so on. According to the types of interference, the interference can be modelled as random interference and deterministic interference.

Random interference includes the noise-like interference and the signal-dependent interference. For the noise-like interference, the noise jammer masks the radar system by producing noise-like signals. The radar sensitivity will be reduced since the CFAR threshold increases with the increase of noise level [1]. In [2], the covariance matrix of the data under test is modified to include a rank-one positive semidefinite matrix due to the presence of noise-like interference. The GLRT decision rule is used and a noise interference detection algorithm is derived. For the signal-dependent interference, the interference and the target share the same subspace under the assumption that the beam of the detection system is narrow enough. In [3], the signal-dependent interference model wherein the dependence of the interference echoes on the transmitted signal is accounted for is exploited and a polarimetric detector is derived.

DOI: 10.1201/9781003477907-6

Deterministic interference usually refers to coherent interference such as deception interference. Since it is usually constrained to lie in a known subspace, coherent interference is also called subspace interference. In [4], one-step GLRT and two-step GLRT are used to solve the problem of detecting distributed targets in homogeneous and partially homogeneous noise plus subspace interference. In [5], the gradient test, Durbin test, and their two-step variants are derived to detect a multichannel subspace signal in the presence of deterministic interference and structure nonhomogeneity.

In this chapter, the problem of adaptive detection of a subspace signal in interference is considered. In Section 6.1, adaptive detectors based on the two-step GLRT, Rao test, and Wald test are designed for a subspace signal in the signal-dependent interference. In Section 6.2 and Section 6.3, the problems of detecting a subspace signal in the subspace interference are considered when the clutter is homogeneous and partially homogeneous, respectively.

6.1 ADAPTIVE DETECTORS IN SIGNAL-DEPENDENT INTERFERENCE

6.1.1 Problem Formulation

The receiver antenna array is assumed to have a narrow beam so that only the interference in the direction of the beam is received. The received data include the target echoes and the undesired reflections from the environment. We denote the primary data and training data by $\boldsymbol{x} \in \mathbb{C}^{N\times 1}$ and $\boldsymbol{x}_t \in \mathbb{C}^{N\times 1}, t = 1,\ldots,K$ and formulate the detection problem as the following binary hypotheses test

$$\begin{cases} H_0 : \begin{cases} \boldsymbol{x} = \boldsymbol{H}\boldsymbol{c} + \boldsymbol{n}, & \\ \boldsymbol{x}_t = \boldsymbol{H}\boldsymbol{c}_t + \boldsymbol{n}_t, & t = 1,\ldots,K, \end{cases} \\ H_1 : \begin{cases} \boldsymbol{x} = \boldsymbol{H}\boldsymbol{a} + \boldsymbol{H}\boldsymbol{c} + \boldsymbol{n}, & \\ \boldsymbol{x}_t = \boldsymbol{H}\boldsymbol{c}_t + \boldsymbol{n}_t, & t = 1,\ldots,K, \end{cases} \end{cases} \tag{6.1}$$

where $\boldsymbol{H} \in \mathbb{C}^{N\times p}$ denotes the subspace of signal and interference, $\text{rank}(\boldsymbol{H}) = p$, $N > p$, $\boldsymbol{a} \in \mathbb{C}^{p\times 1}$ denotes the unknown coordinate vector, $\boldsymbol{c} \in \mathbb{C}^{p\times 1}$ and $\boldsymbol{c}_t \in \mathbb{C}^{p\times 1}, t = 1,\ldots,K$ denote the complex coefficient vectors of the interference and satisfy: $\boldsymbol{c}, \boldsymbol{c}_t \sim \mathcal{CN}_N\left(\boldsymbol{0}, \boldsymbol{\Sigma}\right)$, the additive thermal noise $\boldsymbol{n} \in \mathbb{C}^{N\times 1}$ and $\boldsymbol{n}_t \in \mathbb{C}^{N\times 1}, t = 1,\ldots,K$, which is statistically independent of the interference, satisfy: $\boldsymbol{n}, \boldsymbol{n}_t \sim \mathcal{CN}_N\left(\boldsymbol{0}, \sigma^2 \boldsymbol{I}_N\right)$. We assume that $\boldsymbol{H}$ and σ^2 are known since the noise power can be estimated by averaging the energy of the samples received long before the signal arrival time [6].

From Equation 6.1, the joint PDF for all the data under H_i and the logarithm of the joint PDF are given by

$$f\left(\boldsymbol{x},\boldsymbol{x}_1,\ldots\boldsymbol{x}_K \mid H_i\right)=\frac{1}{\pi^{N(K+1)}\det^{K+1}\left(\boldsymbol{M}\right)} \exp\left\{-\mathrm{tr}\left[\boldsymbol{M}^{-1}\left(\left(\boldsymbol{x}-i\boldsymbol{Ha}\right)\left(\boldsymbol{x}-i\boldsymbol{Ha}\right)^H+\boldsymbol{S}\right)\right]\right\}, \tag{6.2}$$

$$\ln f\left(\boldsymbol{x},\boldsymbol{x}_1,\ldots\boldsymbol{x}_K \mid H_i\right)=-N\left(K+1\right)\ln\pi-\left(K+1\right)\ln\det\left(\boldsymbol{M}\right) -\mathrm{tr}\left\{\boldsymbol{M}^{-1}\left[\left(\boldsymbol{x}-i\boldsymbol{Ha}\right)\left(\boldsymbol{x}-i\boldsymbol{Ha}\right)^H+\boldsymbol{S}\right]\right\}. \tag{6.3}$$

where $i=0$ denotes the hypothesis H_0, $i=1$ denotes the hypothesis H_1, $\boldsymbol{M}=\boldsymbol{H\Sigma H}^H+\sigma^2\boldsymbol{I}_N$, $\boldsymbol{S}=\sum_{t=1}^{K}\boldsymbol{x}_t\boldsymbol{x}_t^H$ is the sample covariance matrix.

According to the Woodbury formula [7] and the determinant identity $\det\left(\boldsymbol{I}_m+\boldsymbol{XY}\right)=\det\left(\boldsymbol{I}_n+\boldsymbol{YX}\right)$ for $\boldsymbol{X}\in\mathbb{C}^{m\times n},\boldsymbol{Y}\in\mathbb{C}^{n\times m}$, the following results can be obtained

$$\boldsymbol{M}^{-1}=\left(\boldsymbol{H\Sigma H}^H+\sigma^2\boldsymbol{I}_N\right)^{-1}=\sigma^{-2}\left[\boldsymbol{I}_N-\boldsymbol{H}\left(\sigma^2\boldsymbol{I}_p+\boldsymbol{\Sigma H}^H\boldsymbol{H}\right)^{-1}\boldsymbol{\Sigma H}^H\right], \tag{6.4}$$

$$\begin{aligned}\det\left(\boldsymbol{M}\right)&=\det\left(\sigma^2\boldsymbol{I}_N+\boldsymbol{H\Sigma H}^H\right)\\&=\sigma^{2N}\det\left(\boldsymbol{I}_p+\sigma^{-2}\boldsymbol{\Sigma H}^H\boldsymbol{H}\right)=\sigma^{2N-2p}\det\left(\sigma^2\boldsymbol{I}_p+\boldsymbol{\Sigma H}^H\boldsymbol{H}\right).\end{aligned} \tag{6.5}$$

$$\begin{aligned}
&\operatorname{tr}\left\{\boldsymbol{M}^{-1}\left[(\boldsymbol{x}-i\boldsymbol{H}\boldsymbol{a})(\boldsymbol{x}-i\boldsymbol{H}\boldsymbol{a})^{H}+\boldsymbol{S}\right]\right\} \\
&=\operatorname{tr}\left\{\sigma^{-2}\left[\boldsymbol{I}_N-\boldsymbol{H}\left(\sigma^2\boldsymbol{I}_p+\boldsymbol{\Sigma}\boldsymbol{H}^H\boldsymbol{H}\right)^{-1}\boldsymbol{\Sigma}\boldsymbol{H}^H\right]\boldsymbol{T}_i\right\} \\
&=\operatorname{tr}\left(\sigma^{-2}\boldsymbol{T}_i\right)-\sigma^{-2}\operatorname{tr}\left[\left(\sigma^2\boldsymbol{I}_p+\boldsymbol{\Sigma}\boldsymbol{H}^H\boldsymbol{H}\right)^{-1}\boldsymbol{\Sigma}\boldsymbol{H}^H\boldsymbol{T}_i\boldsymbol{H}\right] \\
&=\operatorname{tr}\left(\sigma^{-2}\boldsymbol{T}_i\right)-\sigma^{-2}\operatorname{tr}\left[\left(\sigma^2\boldsymbol{I}_p+\boldsymbol{\Sigma}\boldsymbol{H}^H\boldsymbol{H}\right)^{-1}\boldsymbol{\Sigma}\boldsymbol{H}^H\boldsymbol{T}_i\boldsymbol{H}\right. \\
&\qquad \left.+\boldsymbol{\Psi}_H\boldsymbol{T}_i\boldsymbol{H}-\boldsymbol{\Psi}_H\boldsymbol{T}_i\boldsymbol{H}\right] \\
&=\operatorname{tr}\left(\sigma^{-2}\boldsymbol{T}_i\right)-\sigma^{-2}\operatorname{tr}\left(\boldsymbol{\Psi}_H\boldsymbol{T}_i\boldsymbol{H}\right)-\sigma^{-2}\operatorname{tr}\left\{\left(\sigma^2\boldsymbol{I}_p+\boldsymbol{\Sigma}\boldsymbol{H}^H\boldsymbol{H}\right)^{-1}\right. \\
&\qquad \left.\times\left[\boldsymbol{\Sigma}\boldsymbol{H}^H\boldsymbol{T}_i\boldsymbol{H}-\left(\sigma^2\boldsymbol{I}_p+\boldsymbol{\Sigma}\boldsymbol{H}^H\boldsymbol{H}\right)\boldsymbol{\Psi}_H\boldsymbol{T}_i\boldsymbol{H}\right]\right\} \\
&=\operatorname{tr}\left(\sigma^{-2}\boldsymbol{\Psi}_H^{\perp}\boldsymbol{T}_i\right)+\sigma^{-2}\operatorname{tr}\left[\left(\sigma^2\boldsymbol{I}_p+\boldsymbol{\Sigma}\boldsymbol{H}^H\boldsymbol{H}\right)^{-1}\sigma^2\boldsymbol{\Psi}_H\boldsymbol{T}_i\boldsymbol{H}\right] \\
&=\operatorname{tr}\left(\sigma^{-2}\boldsymbol{\Psi}_H^{\perp}\boldsymbol{T}_i\right)+\operatorname{tr}\left\{\left(\boldsymbol{H}^H\boldsymbol{H}\right)^{-1}\left[\sigma^2\left(\boldsymbol{H}^H\boldsymbol{H}\right)^{-1}+\boldsymbol{\Sigma}\right]^{-1}\boldsymbol{\Psi}_H\boldsymbol{T}_i\boldsymbol{H}\right\} \\
&=\operatorname{tr}\left(\sigma^{-2}\boldsymbol{\Psi}_H^{\perp}\boldsymbol{T}_i\right)+\operatorname{tr}\left\{\left[\sigma^2\left(\boldsymbol{H}^H\boldsymbol{H}\right)^{-1}+\boldsymbol{\Sigma}\right]^{-1}\boldsymbol{\Psi}_H\boldsymbol{T}_i\boldsymbol{\Psi}_H^H\right\},
\end{aligned} \tag{6.6}$$

where $\boldsymbol{\Psi}_H=\left(\boldsymbol{H}^H\boldsymbol{H}\right)^{-1}\boldsymbol{H}^H$, $\boldsymbol{\Psi}_H^{\perp}=\boldsymbol{I}_N-\boldsymbol{H}\boldsymbol{\Psi}_H$, $\boldsymbol{T}_i=(\boldsymbol{x}-i\boldsymbol{H}\boldsymbol{a})(\boldsymbol{x}-i\boldsymbol{H}\boldsymbol{a})^H+\boldsymbol{S}$. The equation $\operatorname{tr}(\boldsymbol{XY})=\operatorname{tr}(\boldsymbol{YX})$ has been applied repetitiously in Equation 6.6. We substitute Equations 6.4–6.6 into Equation 6.2 and rewrite the logarithm of the PDF as

$$\begin{aligned}
&\ln f\left(\boldsymbol{x},\boldsymbol{x}_1,\ldots\boldsymbol{x}_K \mid H_i\right)=-N(K+1)\ln\pi \\
&-(K+1)\ln\sigma^{2N-2p}\det\left(\sigma^2\boldsymbol{I}_p+\boldsymbol{\Sigma}\boldsymbol{H}^H\boldsymbol{H}\right) \\
&-\operatorname{tr}\left(\sigma^{-2}\boldsymbol{\Psi}_H^{\perp}\boldsymbol{T}_i\right)-\operatorname{tr}\left\{\left[\sigma^2\left(\boldsymbol{H}^H\boldsymbol{H}\right)^{-1}+\boldsymbol{\Sigma}\right]^{-1}\boldsymbol{\Psi}_H\boldsymbol{T}_i\boldsymbol{\Psi}_H^H\right\}.
\end{aligned} \tag{6.7}$$

6.1.2 The Rao Test

The Rao test for the complex-valued signal is given by [8]

$$\left.\frac{\partial \ln f(\boldsymbol{x},\boldsymbol{x}_1,\ldots\boldsymbol{x}_K \mid \boldsymbol{\theta})}{\partial \boldsymbol{\theta}_r}\right|^T_{\boldsymbol{\theta}=\hat{\boldsymbol{\theta}}_0}\left[\boldsymbol{J}^{-1}(\hat{\boldsymbol{\theta}}_0)\right]_{\boldsymbol{\theta}_r,\boldsymbol{\theta}_r}\times\left.\frac{\partial \ln f(\boldsymbol{x},\boldsymbol{x}_1,\ldots\boldsymbol{x}_K \mid \boldsymbol{\theta})}{\partial \boldsymbol{\theta}_r^*}\right|_{\boldsymbol{\theta}=\hat{\boldsymbol{\theta}}_0}\underset{H_0}{\overset{H_1}{\gtrless}}\xi_1 \tag{6.8}$$

where $\boldsymbol{\theta}=\left[\boldsymbol{\theta}_r^T,\boldsymbol{\theta}_s^T\right]^T\in\mathbb{C}^{(p+p^2)\times 1}$, $\hat{\boldsymbol{\theta}}_0=\left[\hat{\boldsymbol{\theta}}_{r,0}^T,\hat{\boldsymbol{\theta}}_{s,0}^T\right]^T$ is the MLE of $\boldsymbol{\theta}$ under hypothesis H_0, $\boldsymbol{\theta}_r=\boldsymbol{a}\in\mathbb{C}^{p\times 1}$, $\partial/\partial\boldsymbol{\theta}_r$ denotes the gradient with respect to $\boldsymbol{\theta}_r$, $\boldsymbol{\theta}_s=\mathrm{vec}(\boldsymbol{\Sigma})\in\mathbb{C}^{p^2\times 1}$, $\boldsymbol{J}(\boldsymbol{\theta})=\boldsymbol{J}(\boldsymbol{\theta}_r,\boldsymbol{\theta}_s)$ denotes the FIM, and ξ_1 is the detection threshold.

In order to obtain the Rao test, the derivative of the logarithm of PDF $\ln f(\boldsymbol{x},\boldsymbol{x}_1,\ldots\boldsymbol{x}_K \mid \boldsymbol{\theta})$ with respect to $\boldsymbol{\theta}_r$ (namely, $\boldsymbol{a}$) is needed. By using the fact that $\boldsymbol{\Psi}_{\boldsymbol{H}}^{\perp}\boldsymbol{H}=\boldsymbol{0}$ and $\left(\boldsymbol{\Psi}_{\boldsymbol{H}}^{\perp}\right)^H=\boldsymbol{\Psi}_{\boldsymbol{H}}^{\perp}$, $\mathrm{tr}\left(\boldsymbol{\Psi}_{\boldsymbol{H}}^{\perp}\boldsymbol{T}_1\right)$ can be simplified as

$$\begin{aligned}
&\mathrm{tr}\left(\boldsymbol{\Psi}_{\boldsymbol{H}}^{\perp}\boldsymbol{T}_1\right)\\
&=\mathrm{tr}\left[\boldsymbol{\Psi}_{\boldsymbol{H}}^{\perp}\left(\boldsymbol{x}\boldsymbol{x}^H+\boldsymbol{S}+\boldsymbol{H}\boldsymbol{a}\boldsymbol{a}^H\boldsymbol{H}^H-\boldsymbol{x}\boldsymbol{a}^H\boldsymbol{H}^H-\boldsymbol{H}\boldsymbol{a}\boldsymbol{x}^H\right)\right]\\
&=\mathrm{tr}\left[\boldsymbol{\Psi}_{\boldsymbol{H}}^{\perp}\left(\boldsymbol{x}\boldsymbol{x}^H+\boldsymbol{S}\right)\right]\\
&=\mathrm{tr}\left(\boldsymbol{\Psi}_{\boldsymbol{H}}^{\perp}\boldsymbol{T}_0\right).
\end{aligned} \tag{6.9}$$

Thus, the differential of the logarithm of PDF $\ln f(\boldsymbol{x},\boldsymbol{x}_1,\ldots\boldsymbol{x}_K \mid \boldsymbol{\theta},H_1)$ in Equation 6.7 with respect to $\boldsymbol{a}$ is given by

$$\begin{aligned}
&d_{\boldsymbol{a}}\left[\ln f(\boldsymbol{x},\boldsymbol{x}_1,\ldots\boldsymbol{x}_K \mid \boldsymbol{\theta},H_1)\right]\\
&=-d_{\boldsymbol{a}}\mathrm{tr}\left\{\left[\sigma^2\left(\boldsymbol{H}^H\boldsymbol{H}\right)^{-1}+\Sigma\right]^{-1}\boldsymbol{\Psi}_{\boldsymbol{H}}\right.\\
&\left.\times\left(\boldsymbol{x}\boldsymbol{x}^H-\boldsymbol{x}\boldsymbol{a}^H\boldsymbol{H}^H-\boldsymbol{H}\boldsymbol{a}\boldsymbol{x}^H+\boldsymbol{H}\boldsymbol{a}\boldsymbol{a}^H\boldsymbol{H}^H+\boldsymbol{S}\right)\boldsymbol{\Psi}_{\boldsymbol{H}}^H\right\}.
\end{aligned} \tag{6.10}$$

The partial derivatives of each term in the logarithm of PDF with respect to $\boldsymbol{a}^T$ and $\boldsymbol{a}^*$ can be calculated as follows

$$\frac{\partial \mathrm{tr}\left\{\left[\sigma^2\left(\boldsymbol{H}^H\boldsymbol{H}\right)^{-1}+\Sigma\right]^{-1}\Psi_H\boldsymbol{H}\boldsymbol{a}\boldsymbol{a}^H\boldsymbol{H}^H\Psi_H^H\right\}}{\partial\boldsymbol{a}^T}=\boldsymbol{a}^H\boldsymbol{H}^H\Psi_H^H\left[\sigma^2\left(\boldsymbol{H}^H\boldsymbol{H}\right)^{-1}+\Sigma\right]^{-1}\Psi_H\boldsymbol{H}, \tag{6.11}$$

$$\frac{\partial \mathrm{tr}\left\{\left[\sigma^2\left(\boldsymbol{H}^H\boldsymbol{H}\right)^{-1}+\Sigma\right]^{-1}\Psi_H\boldsymbol{H}\boldsymbol{a}\boldsymbol{a}^H\boldsymbol{H}^H\Psi_H^H\right\}}{\partial\boldsymbol{a}^*}=\boldsymbol{H}^H\Psi_H^H\left[\sigma^2\left(\boldsymbol{H}^H\boldsymbol{H}\right)^{-1}+\Sigma\right]^{-1}\Psi_H\boldsymbol{H}\boldsymbol{a}, \tag{6.12}$$

$$\frac{\partial \mathrm{tr}\left\{\left[\sigma^2\left(\boldsymbol{H}^H\boldsymbol{H}\right)^{-1}+\Sigma\right]^{-1}\Psi_H\boldsymbol{x}\boldsymbol{a}^H\boldsymbol{H}^H\Psi_H^H\right\}}{\partial\boldsymbol{a}^T}=\boldsymbol{0}, \tag{6.13}$$

$$\frac{\partial \mathrm{tr}\left\{\left[\sigma^2\left(\boldsymbol{H}^H\boldsymbol{H}\right)^{-1}+\Sigma\right]^{-1}\Psi_H\boldsymbol{x}\boldsymbol{a}^H\boldsymbol{H}^H\Psi_H^H\right\}}{\partial\boldsymbol{a}^*}=\boldsymbol{H}^H\Psi_H^H\left[\sigma^2\left(\boldsymbol{H}^H\boldsymbol{H}\right)^{-1}+\Sigma\right]^{-1}\Psi_H\boldsymbol{x}, \tag{6.14}$$

$$\frac{\partial \mathrm{tr}\left\{\left[\sigma^2\left(\boldsymbol{H}^H\boldsymbol{H}\right)^{-1}+\Sigma\right]^{-1}\Psi_H\boldsymbol{H}\boldsymbol{a}\boldsymbol{x}^H\Psi_H^H\right\}}{\partial\boldsymbol{a}^T}=\boldsymbol{x}^H\Psi_H^H\left[\sigma^2\left(\boldsymbol{H}^H\boldsymbol{H}\right)^{-1}+\Sigma\right]^{-1}\Psi_H\boldsymbol{H}, \tag{6.15}$$

$$\frac{\partial \mathrm{tr}\left\{\left[\sigma^2\left(\boldsymbol{H}^H\boldsymbol{H}\right)^{-1}+\Sigma\right]^{-1}\boldsymbol{\Psi}_H\boldsymbol{H}\boldsymbol{a}\boldsymbol{x}^H\boldsymbol{\Psi}_H^H\right\}}{\partial \boldsymbol{a}^*}=\boldsymbol{0}. \tag{6.16}$$

Note that $\boldsymbol{\Psi}_H\boldsymbol{H}=\boldsymbol{H}^H\boldsymbol{\Psi}_H^H=\boldsymbol{I}_p$. As a consequence, the partial derivatives of the logarithm of PDF $\ln f\left(\boldsymbol{x},\boldsymbol{x}_1,\ldots\boldsymbol{x}_K \mid \boldsymbol{\theta},H_1\right)$ with respect to $\boldsymbol{a}^T$ and $\boldsymbol{a}^*$ can be obtained as

$$\frac{\partial \ln f\left(\boldsymbol{x},\boldsymbol{x}_1,\ldots\boldsymbol{x}_K \mid \boldsymbol{\theta}\right)}{\partial \boldsymbol{a}^T}=\boldsymbol{x}^H\boldsymbol{\Psi}_H^H\left[\sigma^2\left(\boldsymbol{H}^H\boldsymbol{H}\right)^{-1}+\Sigma\right]^{-1}-\boldsymbol{a}^H\left[\sigma^2\left(\boldsymbol{H}^H\boldsymbol{H}\right)^{-1}+\Sigma\right]^{-1}, \tag{6.17}$$

$$\frac{\partial \ln f\left(\boldsymbol{x},\boldsymbol{x}_1,\ldots\boldsymbol{x}_K \mid \boldsymbol{\theta}\right)}{\partial \boldsymbol{a}^*}=\left[\sigma^2\left(\boldsymbol{H}^H\boldsymbol{H}\right)^{-1}+\Sigma\right]^{-1}\boldsymbol{\Psi}_H\boldsymbol{x}-\left[\sigma^2\left(\boldsymbol{H}^H\boldsymbol{H}\right)^{-1}+\Sigma\right]^{-1}\boldsymbol{\Psi}_H\boldsymbol{H}\boldsymbol{a}. \tag{6.18}$$

Next, the FIM will be calculated. The FIM $\boldsymbol{J}(\boldsymbol{\theta})$ can be partitioned as [9] $\boldsymbol{J}(\boldsymbol{\theta})=\begin{bmatrix}\boldsymbol{J}_{\theta_r,\theta_r}(\boldsymbol{\theta}) & \boldsymbol{J}_{\theta_r,\theta_s}(\boldsymbol{\theta})\\ \boldsymbol{J}_{\theta_s,\theta_r}(\boldsymbol{\theta}) & \boldsymbol{J}_{\theta_s,\theta_s}(\boldsymbol{\theta})\end{bmatrix}$, where

$$\boldsymbol{J}_{\theta_r,\theta_r}(\boldsymbol{\theta})=E\left[\frac{\partial \ln f\left(\boldsymbol{x},\boldsymbol{x}_1,\ldots\boldsymbol{x}_K \mid \boldsymbol{\theta}\right)}{\partial \boldsymbol{\theta}_r^*}\frac{\partial \ln f\left(\boldsymbol{x},\boldsymbol{x}_1,\ldots\boldsymbol{x}_K \mid \boldsymbol{\theta}\right)}{\partial \boldsymbol{\theta}_r^T}\right],$$

$$\boldsymbol{J}_{\theta_r,\theta_s}(\boldsymbol{\theta})=E\left[\frac{\partial \ln f\left(\boldsymbol{x},\boldsymbol{x}_1,\ldots\boldsymbol{x}_K \mid \boldsymbol{\theta}\right)}{\partial \boldsymbol{\theta}_r^*}\frac{\partial \ln f\left(\boldsymbol{x},\boldsymbol{x}_1,\ldots\boldsymbol{x}_K \mid \boldsymbol{\theta}\right)}{\partial \boldsymbol{\theta}_s^T}\right], \tag{6.19}$$

$$\boldsymbol{J}_{\theta_s,\theta_r}(\boldsymbol{\theta})=E\left[\frac{\partial \ln f\left(\boldsymbol{x},\boldsymbol{x}_1,\ldots\boldsymbol{x}_K \mid \boldsymbol{\theta}\right)}{\partial \boldsymbol{\theta}_s^*}\frac{\partial \ln f\left(\boldsymbol{x},\boldsymbol{x}_1,\ldots\boldsymbol{x}_K \mid \boldsymbol{\theta}\right)}{\partial \boldsymbol{\theta}_r^T}\right],$$

$$\boldsymbol{J}_{\theta_s,\theta_s}(\boldsymbol{\theta})=E\left[\frac{\partial \ln f\left(\boldsymbol{x},\boldsymbol{x}_1,\ldots\boldsymbol{x}_K \mid \boldsymbol{\theta}\right)}{\partial \boldsymbol{\theta}_s^*}\frac{\partial \ln f\left(\boldsymbol{x},\boldsymbol{x}_1,\ldots\boldsymbol{x}_K \mid \boldsymbol{\theta}\right)}{\partial \boldsymbol{\theta}_s^T}\right].$$

Taking the derivative of Equation 6.18 with respect to $\boldsymbol{a}^T$, yields

$$\frac{\partial^2 \ln f\left(\boldsymbol{x},\boldsymbol{x}_1,\ldots\boldsymbol{x}_K \mid \boldsymbol{\theta}\right)}{\partial \boldsymbol{a}^*\partial \boldsymbol{a}^T}=-\left[\sigma^2\left(\boldsymbol{H}^H\boldsymbol{H}\right)^{-1}+\Sigma\right]^{-1}. \tag{6.20}$$

The differential of (6.18) with respect to Σ is given as

$$d_\Sigma\left[\sigma^2\left(\boldsymbol{H}^H\boldsymbol{H}\right)^{-1}+\Sigma\right]^{-1}\Psi_H\left(\boldsymbol{x}-\boldsymbol{H}\boldsymbol{a}\right)$$

$$=-\left[\sigma^2\left(\boldsymbol{H}^H\boldsymbol{H}\right)^{-1}+\Sigma\right]^{-1}d_\Sigma\Sigma \tag{6.21}$$

$$\times\left[\sigma^2\left(\boldsymbol{H}^H\boldsymbol{H}\right)^{-1}+\Sigma\right]^{-1}\Psi_H\left(\boldsymbol{x}-\boldsymbol{H}\boldsymbol{a}\right).$$

Then, the partial derivative of (6.18) with respect to Σ [10] can be obtained:

$$\frac{\partial^2\ln f\left(\boldsymbol{x},\boldsymbol{x}_1,\ldots\boldsymbol{x}_K\mid\boldsymbol{\theta}\right)}{\partial\boldsymbol{a}^*\partial\Sigma} \tag{6.22}$$

$$=-\left\{\left[\sigma^2\left(\boldsymbol{H}^H\boldsymbol{H}\right)^{-1}+\Sigma\right]^{-1}\Psi_H\left(\boldsymbol{x}-\boldsymbol{H}\boldsymbol{a}\right)\right\}^T\otimes\left[\sigma^2\left(\boldsymbol{H}^H\boldsymbol{H}\right)^{-1}+\Sigma\right]^{-1}.$$

Substituting (6.20) and (6.22) into (6.19) leads to the elements of the FIM:

$$\boldsymbol{J}_{\theta_r,\theta_r}\left(\boldsymbol{\theta}\right)=\left[\sigma^2\left(\boldsymbol{H}^H\boldsymbol{H}\right)^{-1}+\Sigma\right]^{-1}, \tag{6.23}$$

$$\boldsymbol{J}_{\theta_r,\theta_s}\left(\boldsymbol{\theta}\right)=\boldsymbol{0}. \tag{6.24}$$

The term $\left[\boldsymbol{J}^{-1}\left(\boldsymbol{\theta}\right)\right]_{\theta_r,\theta_r}$ in (6.8) is the inverse of $\left\{\left[\boldsymbol{J}^{-1}\left(\boldsymbol{\theta}\right)\right]_{\theta_r,\theta_r}\right\}^{-1}$ which is the Schur complement of $\boldsymbol{J}_{\theta_s,\theta_s}\left(\boldsymbol{\theta}\right)$, namely

$$\left[\boldsymbol{J}^{-1}\left(\boldsymbol{\theta}\right)\right]_{\theta_r,\theta_r}=\left[\boldsymbol{J}_{\theta_r,\theta_r}\left(\boldsymbol{\theta}\right)-\boldsymbol{J}_{\theta_r,\theta_s}\left(\boldsymbol{\theta}\right)\boldsymbol{J}_{\theta_s,\theta_s}^{-1}\left(\boldsymbol{\theta}\right)\boldsymbol{J}_{\theta_s,\theta_r}\left(\boldsymbol{\theta}\right)\right]^{-1}. \tag{6.25}$$

Since the term $\boldsymbol{J}_{\theta_r,\theta_s}\left(\boldsymbol{\theta}\right)$ is null, $\left[\boldsymbol{J}^{-1}\left(\boldsymbol{\theta}\right)\right]_{\theta_r,\theta_r}$ can be simplified as

$$\left[\boldsymbol{J}^{-1}\left(\boldsymbol{\theta}\right)\right]_{\theta_r,\theta_r}=\boldsymbol{J}_{\theta_r,\theta_r}^{-1}\left(\boldsymbol{\theta}\right)=\sigma^2\left(\boldsymbol{H}^H\boldsymbol{H}\right)^{-1}+\Sigma. \tag{6.26}$$

From (6.8), the MLE of $\boldsymbol{\theta}$ under H_0 (i.e.,$\hat{\theta}_0$) is also required to obtain the Rao test. According to the knowledge of the complex-valued matrix differentials of matrix determinant and trace of the matrix [9] [10], the differential of the logarithm of PDF $\ln f\left(\boldsymbol{x},\boldsymbol{x}_1,\ldots\boldsymbol{x}_K \mid \boldsymbol{\theta},H_i\right)$ in (6.7) with respect to Σ is given by

$$
\begin{aligned}
& d_{\Sigma} \ln f\left(\boldsymbol{x},\boldsymbol{x}_1,\ldots\boldsymbol{x}_K \mid \boldsymbol{\theta},H_i\right) \\
& =-(K+1)\operatorname{tr}\left[\left(\sigma^2\boldsymbol{I}_p+\boldsymbol{\Sigma}\boldsymbol{H}^H\boldsymbol{H}\right)^{-1} d_{\Sigma}\left(\sigma^2\boldsymbol{I}_p+\boldsymbol{\Sigma}\boldsymbol{H}^H\boldsymbol{H}\right)\right] \\
& +\operatorname{tr}\left\{\left[\sigma^2\left(\boldsymbol{H}^H\boldsymbol{H}\right)^{-1}+\Sigma\right]^{-1}\boldsymbol{\Psi}_H\boldsymbol{T}_i\boldsymbol{\Psi}_H^H\right. \\
& \left.\times\left[\sigma^2\left(\boldsymbol{H}^H\boldsymbol{H}\right)^{-1}+\Sigma\right]^{-1} d_{\Sigma}\left[\sigma^2\left(\boldsymbol{H}^H\boldsymbol{H}\right)^{-1}+\Sigma\right]\right\} \\
& =-(K+1)\operatorname{tr}\left\{\left(\boldsymbol{H}^H\boldsymbol{H}\right)^{-1}\left[\sigma^2\left(\boldsymbol{H}^H\boldsymbol{H}\right)^{-1}+\Sigma\right]^{-1} d_{\Sigma}(\Sigma)\left(\boldsymbol{H}^H\boldsymbol{H}\right)\right\} \\
& +\operatorname{tr}\left\{\left[\sigma^2\left(\boldsymbol{H}^H\boldsymbol{H}\right)^{-1}+\Sigma\right]^{-1}\boldsymbol{\Psi}_H\boldsymbol{T}_i\boldsymbol{\Psi}_H^H\left[\sigma^2\left(\boldsymbol{H}^H\boldsymbol{H}\right)^{-1}+\Sigma\right]^{-1} d_{\Sigma}\Sigma\right\} \\
& =-(K+1)\operatorname{tr}\left\{\left[\sigma^2\left(\boldsymbol{H}^H\boldsymbol{H}\right)^{-1}+\Sigma\right]^{-1} d_{\Sigma}(\Sigma)\right\} \\
& +\operatorname{tr}\left\{\left[\sigma^2\left(\boldsymbol{H}^H\boldsymbol{H}\right)^{-1}+\Sigma\right]^{-1}\boldsymbol{\Psi}_H\boldsymbol{T}_i\boldsymbol{\Psi}_H^H\left[\sigma^2\left(\boldsymbol{H}^H\boldsymbol{H}\right)^{-1}+\Sigma\right]^{-1} d_{\Sigma}\Sigma\right\}
\end{aligned}
\tag{6.27}
$$

Then, based on the differential of the logarithm of PDF $\ln f\left(\boldsymbol{x},\boldsymbol{x}_1,\ldots\boldsymbol{x}_K \mid \boldsymbol{\theta},H_i\right)$ with respect to Σ in (6.27), the partial derivative of the logarithm of PDF with respect to Σ can be obtained as

$$
\begin{aligned}
\frac{\partial \ln f\left(\boldsymbol{x},\boldsymbol{x}_1,\ldots\boldsymbol{x}_K \mid \boldsymbol{\theta},H_i\right)}{\partial \Sigma} & =-(K+1)\left[\sigma^2\left(\boldsymbol{H}^H\boldsymbol{H}\right)^{-1}+\Sigma\right]^{-T} \\
& +\left[\sigma^2\left(\boldsymbol{H}^H\boldsymbol{H}\right)^{-1}+\Sigma\right]^{-T}\left(\boldsymbol{\Psi}_H\boldsymbol{T}_i\boldsymbol{\Psi}_H^H\right)^T\left[\sigma^2\left(\boldsymbol{H}^H\boldsymbol{H}\right)^{-1}+\Sigma\right]^{-T}.
\end{aligned}
\tag{6.28}
$$

Setting (6.28) to be zero and setting $i = 0$ yields the MLE of Σ under hypothesis H_0

$$\hat{\Sigma}_0 = \left(\boldsymbol{\Psi}_{\boldsymbol{H}} \boldsymbol{T}_0 \boldsymbol{\Psi}_{\boldsymbol{H}}^H\right) / (K+1) - \sigma^2 \left(\boldsymbol{H}^H \boldsymbol{H}\right)^{-1}. \tag{6.29}$$

The MLE of $\boldsymbol{\theta}_r$ under hypothesis H_0 is found to be

$$\hat{\boldsymbol{\theta}}_{r,0} = \boldsymbol{\theta}_{r,0} = \boldsymbol{0}_{p\times 1}. \tag{6.30}$$

Finally, by substituting (6.17)–(6.18), (6.26), and (6.29)–(6.30) into (6.8) and rearranging the expression, the new Rao test can be obtained as

$$\boldsymbol{x}^H \boldsymbol{\Psi}_{\boldsymbol{H}}^H \left[\left(\boldsymbol{\Psi}_{\boldsymbol{H}} \left(\boldsymbol{x}\boldsymbol{x}^H + \boldsymbol{S}\right) \boldsymbol{\Psi}_{\boldsymbol{H}}^H\right) / (K+1)\right]^{-1} \boldsymbol{\Psi}_{\boldsymbol{H}} \boldsymbol{x} \underset{H_0}{\overset{H_1}{\gtrless}} \xi_{SD-Rao} \tag{6.31}$$

where ξ_{SD-Rao} denotes the threshold.

6.1.3 The Wald Test

The Wald test for the complex-valued signal is given as [8]

$$\hat{\boldsymbol{\theta}}_{r,1}^H \left(\left[\boldsymbol{J}^{-1}\left(\hat{\boldsymbol{\theta}}_1\right)\right]_{\boldsymbol{\theta}_r, \boldsymbol{\theta}_r}\right)^{-1} \hat{\boldsymbol{\theta}}_{r,1} \underset{H_0}{\overset{H_1}{\gtrless}} \xi_2 \tag{6.32}$$

where $\hat{\boldsymbol{\theta}}_{r,1} = \hat{\boldsymbol{a}}_1$ is the MLE of $\boldsymbol{\theta}_r$ under H_1, $\hat{\boldsymbol{\theta}}_{s,1} = \text{vec}\left(\hat{\Sigma}_1\right)$ is the MLE of $\boldsymbol{\theta}_s$ under H_1, $\hat{\boldsymbol{\theta}}_1 = \left[\hat{\boldsymbol{\theta}}_{r,1}^T, \hat{\boldsymbol{\theta}}_{s,1}^T\right]^T$ is the MLE of $\boldsymbol{\theta}$ under H_1, ξ_2 is the detection threshold of the Wald test.

From (6.32), the MLE of $\boldsymbol{\theta}$ under hypothesis H_1 (i.e., $\hat{\boldsymbol{\theta}}_1$) is needed to obtain the new Wald test. Nulling the derivative of $\ln f\left(\boldsymbol{x}, \boldsymbol{x}_1, \ldots \boldsymbol{x}_K \mid \boldsymbol{\theta}\right)$ with respect to $\boldsymbol{\theta}_r$ in (6.17) yields the MLE of $\boldsymbol{\theta}_r$ under hypothesis H_1:

$$\hat{\boldsymbol{\theta}}_{r,1} = \hat{\boldsymbol{a}}_1 = \boldsymbol{\Psi}_{\boldsymbol{H}} \boldsymbol{x}. \tag{6.33}$$

In the same way, the MLE of Σ under hypothesis H_1 can be obtained by equating the partial derivative in (6.28) to zero and setting $i = 1$:

$$\hat{\theta}_{s,1} = \mathrm{vec}\left(\hat{\Sigma}_1\right) = \mathrm{vec}\left[\boldsymbol{\Psi}_H \hat{\boldsymbol{T}}_1 \boldsymbol{\Psi}_H^H / (K+1) - \sigma^2\left(\boldsymbol{H}^H \boldsymbol{H}\right)^{-1}\right], \tag{6.34}$$

where $\hat{\boldsymbol{T}}_1 = \left(\boldsymbol{x} - \boldsymbol{H}\hat{\boldsymbol{a}}_1\right)\left(\boldsymbol{x} - \boldsymbol{H}\hat{\boldsymbol{a}}_1\right)^H + \boldsymbol{S}$. Plugging (6.33) and (6.34) into (6.23) results in

$$\left[\boldsymbol{J}^{-1}\left(\hat{\theta}_1\right)\right]_{\theta_r,\theta_r} = \sigma^2\left(\boldsymbol{H}^H \boldsymbol{H}\right)^{-1} + \hat{\Sigma}_1. \tag{6.35}$$

Finally, by substituting (6.33)–(6.35) into (6.32) and rearranging the expression, a new Wald test can be obtained

$$\boldsymbol{x}^H \boldsymbol{\Psi}_H^H \left[\boldsymbol{\Psi}_H \check{\boldsymbol{T}}_1 \boldsymbol{\Psi}_H^H / (K+1)\right]^{-1} \boldsymbol{\Psi}_H \boldsymbol{x} \underset{H_0}{\overset{H_1}{\gtrless}} \xi_2 \tag{6.36}$$

By using the equality $\boldsymbol{\Psi}_H \boldsymbol{H} = \boldsymbol{H}^H \boldsymbol{\Psi}_H^H = \boldsymbol{I}_p$, $\boldsymbol{\Psi}_H \hat{\boldsymbol{T}}_1 \boldsymbol{\Psi}_H^H$ can be rewritten as follows

$$\begin{aligned} \boldsymbol{\Psi}_H \hat{\boldsymbol{T}}_1 \boldsymbol{\Psi}_H^H &= \boldsymbol{\Psi}_H \left[\left(\boldsymbol{x} - \boldsymbol{H}\hat{\boldsymbol{a}}_1\right)\left(\boldsymbol{x} - \boldsymbol{H}\hat{\boldsymbol{a}}_1\right)^H + \boldsymbol{S}\right] \boldsymbol{\Psi}_H^H \\ &= \boldsymbol{\Psi}_H \boldsymbol{x}\boldsymbol{x}^H \boldsymbol{\Psi}_H^H - \boldsymbol{\Psi}_H \boldsymbol{x}\boldsymbol{x}^H \boldsymbol{\Psi}_H^H \boldsymbol{H}^H \boldsymbol{\Psi}_H^H - \boldsymbol{\Psi}_H \boldsymbol{H} \boldsymbol{\Psi}_H \boldsymbol{x}\boldsymbol{x}^H \boldsymbol{\Psi}_H^H \\ &\quad + \boldsymbol{\Psi}_H \boldsymbol{H} \boldsymbol{\Psi}_H \boldsymbol{x}\boldsymbol{x}^H \boldsymbol{\Psi}_H^H \boldsymbol{H}^H \boldsymbol{\Psi}_H^H + \boldsymbol{\Psi}_H \boldsymbol{S} \boldsymbol{\Psi}_H^H = \boldsymbol{\Psi}_H \boldsymbol{S} \boldsymbol{\Psi}_H^H. \end{aligned} \tag{6.37}$$

Ignoring the constants, after some calculation, (6.36) can be recast as

$$T_{SD\text{-}Wald} = \boldsymbol{x}^H \boldsymbol{\Psi}_H^H \left(\boldsymbol{\Psi}_H \boldsymbol{S} \boldsymbol{\Psi}_H^H\right)^{-1} \boldsymbol{\Psi}_H \boldsymbol{x} \underset{H_0}{\overset{H_1}{\gtrless}} \xi_{SD\text{-}Wald} \tag{6.38}$$

where $\xi_{SD\text{-}Wald}$ denotes the detection threshold.

6.1.4 The Generalized Likelihood Ratio Test

In this subsection, the two-step GLRT criterion is resorted to solve the detection problem in Equation 6.1. More precisely, the interference covariance matrix Σ is first assumed to be known and the GLRT is derived; then, the fully adaptive detector is obtained by replacing Σ with its estimation $\hat{\Sigma}_0$ based on the training data. Under the assumption that the interference covariance matrix is known, the GLRT is given by

$$\frac{\max_{\boldsymbol{a}} f_1(\boldsymbol{x}|\boldsymbol{a})}{f_0(\boldsymbol{x})} \underset{H_0}{\overset{H_1}{\gtrless}} \xi_3 \tag{6.39}$$

where $f_i(\cdot)$ denotes PDF of the primary data under hypothesis $H_i, i=0,1$, given by

$$f_i(\boldsymbol{x}) = \frac{1}{\pi^N \det(\boldsymbol{M})} \times \exp\left\{-\mathrm{tr}\left[\boldsymbol{M}^{-1}(\boldsymbol{x}-i\boldsymbol{Ha})(\boldsymbol{x}-i\boldsymbol{Ha})^H\right]\right\}, \tag{6.40}$$

In Equation 6.40, $\boldsymbol{M} = \boldsymbol{H}\boldsymbol{\Sigma}\boldsymbol{H}^H + \sigma^2\boldsymbol{I}_N$ is the covariance matrix of the received data. After some calculation, the logarithm of the PDF can be rewritten as

$$\begin{aligned} \ln f_i(\boldsymbol{x}) = &-N\ln\pi - \ln\sigma^{2N-2p}\det\left(\sigma^2\boldsymbol{I}_p + \boldsymbol{\Sigma}\boldsymbol{H}^H\boldsymbol{H}\right) \\ &-\mathrm{tr}\left(\sigma^{-2}\boldsymbol{\Psi}_{\boldsymbol{H}}^{\perp}\boldsymbol{U}_i\right) - \mathrm{tr}\left\{\left[\sigma^2\left(\boldsymbol{H}^H\boldsymbol{H}\right)^{-1} + \boldsymbol{\Sigma}\right]^{-1}\boldsymbol{\Psi}_{\boldsymbol{H}}\boldsymbol{U}_i\boldsymbol{\Psi}_{\boldsymbol{H}}^H\right\}. \end{aligned} \tag{6.41}$$

where $\boldsymbol{\Psi}_{\boldsymbol{H}} = \left(\boldsymbol{H}^H\boldsymbol{H}\right)^{-1}\boldsymbol{H}^H$, $\boldsymbol{\Psi}_{\boldsymbol{H}}^{\perp} = \boldsymbol{I}_N - \boldsymbol{H}\boldsymbol{\Psi}_{\boldsymbol{H}}$, and $\boldsymbol{U}_i = (\boldsymbol{x}-i\boldsymbol{Ha})(\boldsymbol{x}-i\boldsymbol{Ha})^H$. Substituting $\boldsymbol{U}_1$ into $\mathrm{tr}\left(\sigma^{-2}\boldsymbol{\Psi}_{\boldsymbol{H}}^{\perp}\boldsymbol{U}_1\right)$, we simplify $\mathrm{tr}\left(\sigma^{-2}\boldsymbol{\Psi}_{\boldsymbol{H}}^{\perp}\boldsymbol{U}_1\right)$ as

$$\mathrm{tr}\left(\sigma^{-2}\boldsymbol{\Psi}_{\boldsymbol{H}}^{\perp}\boldsymbol{U}_1\right) = \sigma^{-2}\mathrm{tr}\left\{\boldsymbol{\Psi}_{\boldsymbol{H}}^{\perp}\left[(\boldsymbol{x}-\boldsymbol{Ha})(\boldsymbol{x}-\boldsymbol{Ha})^H\right]\right\} = \sigma^{-2}\mathrm{tr}\left[\boldsymbol{\Psi}_{\boldsymbol{H}}^{\perp}\boldsymbol{x}\boldsymbol{x}^H\right] \tag{6.42}$$

It is clear to see that $\mathrm{tr}\left(\sigma^{-2}\boldsymbol{\Psi}_{H}^{\perp}\boldsymbol{U}_{1}\right)$ is independent of $\boldsymbol{a}$. Thus, maximizing the PDF under hypothesis H_1 with respect to $\boldsymbol{a}$ is equivalent to solving the following minimization problem

$$\min_{\boldsymbol{a}} \mathrm{tr}\left\{\left[\sigma^2\left(\boldsymbol{H}^H\boldsymbol{H}\right)^{-1}+\boldsymbol{\Sigma}\right]^{-1}\boldsymbol{\Psi}_H\boldsymbol{U}_1\boldsymbol{\Psi}_H^H\right\}$$

$$=\min_{\boldsymbol{a}} \mathrm{tr}\left\{\left[\sigma^2\left(\boldsymbol{H}^H\boldsymbol{H}\right)^{-1}+\boldsymbol{\Sigma}\right]^{-1}\boldsymbol{\Psi}_H\left(\boldsymbol{x}\boldsymbol{x}^H-\boldsymbol{x}\boldsymbol{a}^H\boldsymbol{H}^H-\boldsymbol{H}\boldsymbol{a}\boldsymbol{x}^H+\boldsymbol{H}\boldsymbol{a}\boldsymbol{a}^H\boldsymbol{H}^H\right)\boldsymbol{\Psi}_H^H\right\} \tag{6.43}$$

Taking the partial derivation of $\mathrm{tr}\left\{\left[\sigma^2\left(\boldsymbol{H}^H\boldsymbol{H}\right)^{-1}+\boldsymbol{\Sigma}\right]^{-1}\boldsymbol{\Psi}_H\boldsymbol{U}_1\boldsymbol{\Psi}_H^H\right\}$ with respect to $\boldsymbol{a}^*$ yields

$$\frac{\partial \mathrm{tr}\left\{\left[\sigma^2\left(\boldsymbol{H}^H\boldsymbol{H}\right)^{-1}+\boldsymbol{\Sigma}\right]^{-1}\boldsymbol{\Psi}_H\boldsymbol{U}_1\boldsymbol{\Psi}_H^H\right\}}{\partial\boldsymbol{a}^*} \tag{6.44}$$

$$=\left[\sigma^2\left(\boldsymbol{H}^H\boldsymbol{H}\right)^{-1}+\boldsymbol{\Sigma}\right]^{-1}\boldsymbol{\Psi}_H\boldsymbol{x}-\left[\sigma^2\left(\boldsymbol{H}^H\boldsymbol{H}\right)^{-1}+\boldsymbol{\Sigma}\right]^{-1}\boldsymbol{\Psi}_H\boldsymbol{H}\boldsymbol{a}.$$

The MLE of $\boldsymbol{a}$ can be obtained by nulling the derivation to zero:

$$\hat{\boldsymbol{a}}=\boldsymbol{\Psi}_H\boldsymbol{x} \tag{6.45}$$

Substituting (6.45) into (6.39), the GLRT with known $\boldsymbol{\Sigma}$ is

$$\mathrm{tr}\left\{\left[\sigma^2\left(\boldsymbol{H}^H\boldsymbol{H}\right)^{-1}+\boldsymbol{\Sigma}\right]^{-1}\boldsymbol{\Psi}_H\boldsymbol{x}\boldsymbol{x}^H\boldsymbol{\Psi}_H^H\right\}\underset{H_0}{\overset{H_1}{\gtrless}}\xi_3 \tag{6.46}$$

In the second step, $\boldsymbol{\Sigma}$ is estimated by using the training data. The logarithm of the PDF of the training data is

$$\ln f\left(\boldsymbol{x}_1,\ldots \boldsymbol{x}_K \mid H_i\right) = -NK\ln\pi - K\ln\sigma^{2N-2p}\det\left(\sigma^2\boldsymbol{I}_p + \boldsymbol{\Sigma}\boldsymbol{H}^H\boldsymbol{H}\right)$$
$$-\mathrm{tr}\left(\sigma^{-2}\boldsymbol{\Psi}_H^{\perp}\boldsymbol{S}\right) - \mathrm{tr}\left\{\left[\sigma^2\left(\boldsymbol{H}^H\boldsymbol{H}\right)^{-1} + \Sigma\right]^{-1}\boldsymbol{\Psi}_H\boldsymbol{S}\boldsymbol{\Psi}_H^H\right\}. \tag{6.47}$$

where $S = \sum_{t=1}^{K} x_t x_t^H$ denotes the sample covariance matrix. Taking the differential of the logarithm of the PDF of the training data with respect to Σ yields

$$d_\Sigma \ln f\left(\boldsymbol{x}_1,\ldots \boldsymbol{x}_K\right) = -K\mathrm{tr}\left\{\left[\sigma^2\left(\boldsymbol{H}^H\boldsymbol{H}\right)^{-1} + \Sigma\right]^{-1} d_\Sigma\left(\Sigma\right)\right\}$$
$$+\mathrm{tr}\left\{\left[\sigma^2\left(\boldsymbol{H}^H\boldsymbol{H}\right)^{-1} + \Sigma\right]^{-1}\boldsymbol{\Psi}_H\boldsymbol{S}\boldsymbol{\Psi}_H^H \right. \tag{6.48}$$
$$\left.\times\left[\sigma^2\left(\boldsymbol{H}^H\boldsymbol{H}\right)^{-1} + \Sigma\right]^{-1} d_\Sigma\Sigma\right\}$$

Thus, the partial derivation of Equation 6.47 with respect to Σ is

$$\frac{\partial \ln f\left(\boldsymbol{x}_1,\ldots \boldsymbol{x}_K\right)}{\partial\Sigma} = -K\left[\sigma^2\left(\boldsymbol{H}^H\boldsymbol{H}\right)^{-1} + \Sigma\right]^{-T}$$
$$+\left[\sigma^2\left(\boldsymbol{H}^H\boldsymbol{H}\right)^{-1} + \Sigma\right]^{-T}\left(\boldsymbol{\Psi}_H\boldsymbol{S}\boldsymbol{\Psi}_H^H\right)^T\left[\sigma^2\left(\boldsymbol{H}^H\boldsymbol{H}\right)^{-1} + \Sigma\right]^{-T}. \tag{6.49}$$

The MLE of Σ is obtained by nulling Equation 6.49 to zero:

$$\hat{\Sigma}_0 = \left(\boldsymbol{\Psi}_H\boldsymbol{S}\boldsymbol{\Psi}_H^H\right)/K - \sigma^2\left(\boldsymbol{H}^H\boldsymbol{H}\right)^{-1}. \tag{6.50}$$

The fully adaptive AMF is obtained by plugging Equation 6.50 into Equation 6.46:

$$\mathrm{tr}\left\{\boldsymbol{x}^H\boldsymbol{\Psi}_H^H\left(\boldsymbol{\Psi}_H\boldsymbol{S}\boldsymbol{\Psi}_H^H\right)^{-1}\boldsymbol{\Psi}_H\boldsymbol{x}\right\} \underset{H_0}{\overset{H_1}{\gtrless}} \xi_{SD-AMF} \tag{6.51}$$

It can be seen that the AMF detector is exactly the new Wald test. For convenience, both the new Wald test and AMF are denoted as a new Wald test.

6.1.5 Analytical Performance

We analyze the statistical properties of the designed new Rao test and the new Wald test. An alternative representation of the new Rao test can be obtained by exploiting the matrix inversion lemma [11] :

$$\left[\left(\boldsymbol{\Psi}_H\left(\boldsymbol{x}\boldsymbol{x}^H+\boldsymbol{S}\right)\boldsymbol{\Psi}_H^H\right)/(K+1)\right]^{-1}$$
$$=(K+1)\left[\left(\boldsymbol{\Psi}_H\boldsymbol{S}\boldsymbol{\Psi}_H^H\right)^{-1}-\frac{\left(\boldsymbol{\Psi}_H\boldsymbol{S}\boldsymbol{\Psi}_H^H\right)^{-1}\boldsymbol{\Psi}_H\boldsymbol{x}\boldsymbol{x}^H\boldsymbol{\Psi}_H^H\left(\boldsymbol{\Psi}_H\boldsymbol{S}\boldsymbol{\Psi}_H^H\right)^{-1}}{1+\boldsymbol{x}^H\boldsymbol{\Psi}_H^H\left(\boldsymbol{\Psi}_H\boldsymbol{S}\boldsymbol{\Psi}_H^H\right)^{-1}\boldsymbol{\Psi}_H\boldsymbol{x}}\right]. \tag{6.52}$$

Hence, by ignoring the constants, the new Rao test in Equation 6.31 can be rewritten as

$$T_{SD-Rao}=\boldsymbol{x}^H\boldsymbol{\Psi}_H^H\left[\left(\boldsymbol{\Psi}_H\boldsymbol{S}\boldsymbol{\Psi}_H^H\right)^{-1}-\frac{\left(\boldsymbol{\Psi}_H\boldsymbol{S}\boldsymbol{\Psi}_H^H\right)^{-1}\boldsymbol{\Psi}_H\boldsymbol{x}\boldsymbol{x}^H\boldsymbol{\Psi}_H^H\left(\boldsymbol{\Psi}_H\boldsymbol{S}\boldsymbol{\Psi}_H^H\right)^{-1}}{1+\boldsymbol{x}^H\boldsymbol{\Psi}_H^H\left(\boldsymbol{\Psi}_H\boldsymbol{S}\boldsymbol{\Psi}_H^H\right)^{-1}\boldsymbol{\Psi}_H\boldsymbol{x}}\right]\boldsymbol{\Psi}_H\boldsymbol{x}$$
$$=\frac{\boldsymbol{x}^H\boldsymbol{\Psi}_H^H\left(\boldsymbol{\Psi}_H\boldsymbol{S}\boldsymbol{\Psi}_H^H\right)^{-1}\boldsymbol{\Psi}_H\boldsymbol{x}}{1+\boldsymbol{x}^H\boldsymbol{\Psi}_H^H\left(\boldsymbol{\Psi}_H\boldsymbol{S}\boldsymbol{\Psi}_H^H\right)^{-1}\boldsymbol{\Psi}_H\boldsymbol{x}}\underset{H_0}{\overset{H_1}{\gtrless}}\xi_{SD-Rao} \tag{6.53}$$

where ξ_{SD-Rao} denotes the detection threshold of the new Rao test.

Let $\boldsymbol{y}=\boldsymbol{\Psi}_H\boldsymbol{x}$ and $\breve{\boldsymbol{S}}=\boldsymbol{\Psi}_H\boldsymbol{S}\boldsymbol{\Psi}_H^H=\sum_{t=1}^{K}\boldsymbol{\Psi}_H\boldsymbol{x}_t\boldsymbol{x}_t^H\boldsymbol{\Psi}_H^H$, where $\boldsymbol{x}_t$ follows complex Gaussian distribution with zero mean and covariance matrix $\boldsymbol{H\Sigma H}^H+\sigma^2\boldsymbol{I}_N$ under both hypothesis H_0 and hypothesis H_1,

$\boldsymbol{x}$ is distributed $\mathcal{CN}_N\left(\boldsymbol{0}, \boldsymbol{H}\boldsymbol{\Sigma}\boldsymbol{H}^H + \sigma^2\boldsymbol{I}_N\right)$ under hypothesis H_0 and $\mathcal{CN}_N\left(\boldsymbol{Ha}, \boldsymbol{H}\boldsymbol{\Sigma}\boldsymbol{H}^H + \sigma^2\boldsymbol{I}_N\right)$ under hypothesis H_1. Then, the statistic characters of $\boldsymbol{y}$ can be obtained as

$$\boldsymbol{y} \sim \begin{cases} \mathcal{CN}_N\left(\boldsymbol{0}, \boldsymbol{\Sigma} + \sigma^2\boldsymbol{\Psi}_H\boldsymbol{\Psi}_H^H\right), & H_0 \\ \mathcal{CN}_N\left(\boldsymbol{a}, \boldsymbol{\Sigma} + \sigma^2\boldsymbol{\Psi}_H\boldsymbol{\Psi}_H^H\right), & H_1. \end{cases} \tag{6.54}$$

Denote $\Xi = \boldsymbol{x}^H\boldsymbol{\Psi}_H^H\left(\boldsymbol{\Psi}_H\boldsymbol{S}\boldsymbol{\Psi}_H^H\right)^{-1}\boldsymbol{\Psi}_H\boldsymbol{x} = \boldsymbol{y}^H\breve{\boldsymbol{S}}^{-1}\boldsymbol{y}$. Applying Theorem 5.2.2 in [12], it is verified that the statistic distribution of Ξ is

$$\begin{aligned} H_0 &: \left(\frac{K-p+1}{p}\right)\Xi \sim \mathcal{F}_{2p,2(K-p+1)} \\ H_1 &: \left(\frac{K-p+1}{p}\right)\Xi \sim \mathcal{F}_{2p,2(K-p+1)}(\gamma), \end{aligned} \tag{6.55}$$

where $\mathcal{F}_{2p,2(K-p+1)}$ is the central F distribution with $2p$ and $2(K-p+1)$ degrees of freedom, $\mathcal{F}_{2p,2(K-p+1)}(\gamma)$ is the non-central F distribution with $2p$ and $2(K-p+1)$ degrees of freedom and non-centrality parameterγ. The non-centrality parameter γ is expressed as $\gamma = 2\boldsymbol{a}^H\left(\boldsymbol{\Sigma} + \sigma^2\boldsymbol{\Psi}_H\boldsymbol{\Psi}_H^H\right)^{-1}\boldsymbol{a}$.

Notice that both the test statistic of the new Rao and that of the new Wald are linear transformations of Ξ, i.e., $T_{SD-Rao} = \Xi/(1+\Xi)$ and $T_{SD-Wald} = \Xi$. As a consequence, the PD for the new Rao test is given by

$$\begin{aligned} \mathrm{P}_{d,SD-Rao} &= \mathrm{P}\left(T_{SD-Rao} > \xi_{SD-Rao}\right) \\ &= \mathrm{P}\left(\Xi > \frac{\xi_{SD-Rao}}{1-\xi_{SD-Rao}}\right) = Q_{\mathcal{F}_{2p,2(K-p+1)}(\gamma)}\left(\frac{\xi_{SD-Rao}}{1-\xi_{SD-Rao}}\right), \end{aligned} \tag{6.56}$$

where $Q(\cdot)$ denotes the right-tail probability function. The PFA for the new Rao test is obtained by setting the non-centrality parameter γ to zero, i.e.,

$$\mathrm{P}_{fa,SD-Rao} = Q_{\mathcal{F}_{2p,2(K-p+1)}}\left(\frac{\xi_{SD-Rao}}{1-\xi_{SD-Rao}}\right). \tag{6.57}$$

In a similar way, the PFA and PD of the new Wald test can be obtained as

$$\begin{aligned} P_{fa,SD-Wald} &= Q_{\mathcal{F}_{2p,2(K-p+1)}}\left(\xi_{SD-Wald}\right), \\ P_{d,SD-Wald} &= P\left(T_{SD-Wald} > \xi_{SD-Wald}\right) \\ &= P\left(\Xi > \xi_{SD-Wald}\right) = Q_{\mathcal{F}_{2p,2(K-p+1)}(\gamma)}\left(\xi_{SD-Wald}\right). \end{aligned} \tag{6.58}$$

From Equations 6.57 and 6.58, the new Rao test and the new Wald test have the CFAR property since PFA for the new Rao test and the new Wald test are independent of the covariance matrix of the interference and the noise.

By comparing the new Wald test statistic in Equation 6.38, Equation 6.51, and the new Rao test statistic in Equation 6.53, as well as the PFAs and PDs of the new Rao test and new Wald test, it can be seen that the new Rao test has the same structure as the new Wald test. Thus, for the subspace signal detection problem in the presence of signal-dependent interference and thermal noise, the equivalence between the Rao test and the Wald test also holds for finite data records.

Moreover, it is worth noting the new Rao test and the new Wald test require fewer training data than the conventional detectors. In particular, it is well known that conventional detectors require at least $K = N$ training data to obtain a nonsingular sample covariance matrix $\boldsymbol{S}$. In contrast, for new Wald in Equation 6.38 and new Rao in Equation 6.53, only $K = p$ $(p < N)$ training data are needed to ensure that the term $\left(\boldsymbol{\Psi}_H \boldsymbol{S} \boldsymbol{\Psi}_H^H\right)$ is nonsingular.

6.1.6 Numerical Examples

In this section, Monte Carlo simulations are conducted to investigate the performance of the designed detectors. $100/P_{fa}$ independent Monte Carlo trials are conducted to obtain the thresholds and PDs. The Signal to Interference and Noise Ratio (SINR) and the Interference to Noise Ratio (INR) are defined as [1] $\mathrm{SINR} = \frac{\|\boldsymbol{a}\|^2}{\mathrm{tr}(\Sigma) + \sigma^2}$ and $\mathrm{INR} = \mathrm{tr}(\Sigma)/\sigma^2$. The detection performance of the multidimensional Wald test [13], the GLRT [14], and the Rao test [15] is also given for comparison. The test statistics of the three competitors are:

$$T_{GLRT} = \frac{x^H S^{-1} H \left(H^H S^{-1} H\right)^{-1} H^H S^{-1} x}{1 + x^H S^{-1} x},$$

$$T_{Wald} = x^H S^{-1} H \left(H^H S^{-1} H\right)^{-1} H^H S^{-1} x,$$

$$T_{Rao} = \frac{T_{Wald}}{\left(1 + x^H S^{-1} x\right)\left(1 + x^H S^{-1} x - T_{Wald}\right)}.$$

The analytical results are compared with those from the Monte Carlo trials to verify the analytical results of the new Rao and new Wald tests. In Figure 6.1, the entries of the subspace matrix $\boldsymbol{H}$ and $\boldsymbol{a}$ are drawn from a complex Gaussian distribution with zero mean and unit variance. The covariance matrix is set to be $\Sigma = \rho_c \boldsymbol{I}_p$, where ρ_c is a random number larger than zero. The thermal noise power σ^2 is selected to fulfil the required INR. The curves in Figure 6.1 are averaged results for 100 cases of Σ, $\boldsymbol{H}$, and $\boldsymbol{a}$. The analytical expressions for the PDs of the new Rao and new Wald tests are obtained from (6.56) and (6.58), respectively. The analytical thresholds of the new Rao and the new Wald tests are obtained from (6.57) and (6.58), respectively. The Monte Carlo simulation results of the new Wald and the new Rao tests are obtained from (6.38) and (6.53), respectively.

From Figure 6.1, the results obtained by Monte Carlo trials and the analytical results are coincident. Meanwhile, the new Wald test and the new Rao test are equivalent.

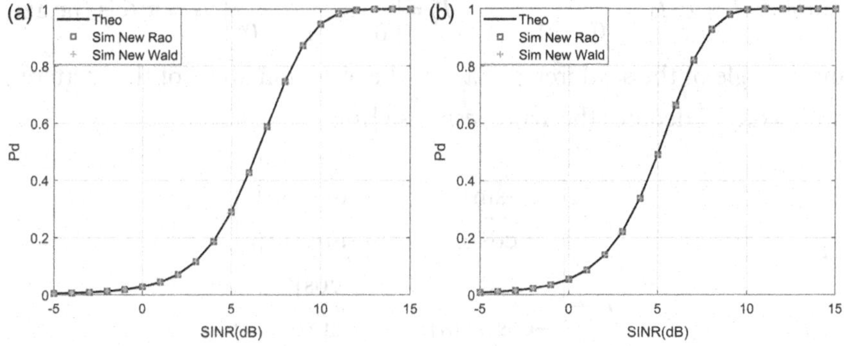

FIGURE 6.1 Probability of detection versus the SINR. (a) $N=6, p=3$; (b) $N=9, p=2$.

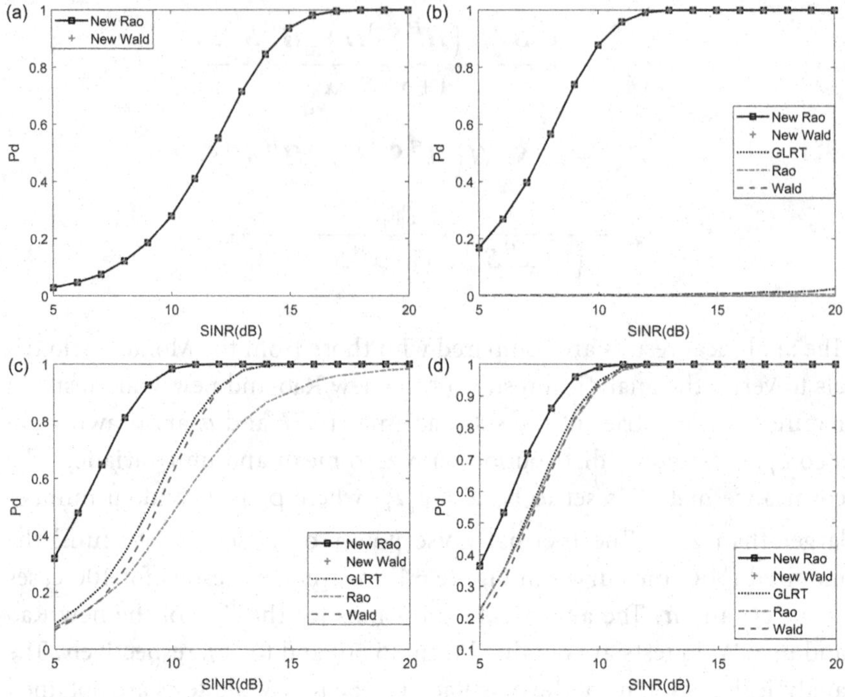

FIGURE 6.2. Probability of detection versus the SINR. (a)$K = 6$;(b)$K = 12$; (c)$K = 24$;(d)$K = 48$.

In Figure 6.2, we set $\boldsymbol{a} = \rho_a[1,1,-0.5j]^T$, $\Sigma = \boldsymbol{I}$, $\boldsymbol{H} = \left[\boldsymbol{h}_1^T,\ldots,\boldsymbol{h}_L^T\right]^T$, where$\boldsymbol{h}_l = \boldsymbol{s} \otimes \boldsymbol{B}\boldsymbol{P}_l$, $\boldsymbol{s} = \left[s_1, s_2, \ldots, s_{N_s}\right]^T$ denotes the sampling instants, $N_s = 1$, $\boldsymbol{P}_l \in C_{2\times p}$,$l = 1,\ldots,L$ denotes the polarization matrix, L denotes the pulse number, $L = 2$, $\boldsymbol{P}_1 = \begin{bmatrix} 1 & 0 & 0 \\ 0 & 0 & 1 \end{bmatrix}$, $\boldsymbol{P}_2 = \begin{bmatrix} 0 & 0 & -1 \\ 0 & -1 & 0 \end{bmatrix}$, $\alpha = \pi/6$ is the azimuth angle of the scatterer, $\beta = \pi/5$ is the elevation angle of the scatterer, and $\boldsymbol{B} \in \mathbb{C}^{T\times 2}$ denotes the array response [16]:

$$\boldsymbol{B} = \begin{bmatrix} -\sin\alpha & -\cos\alpha\sin\beta \\ \cos\alpha & -\sin\alpha\sin\beta \\ 0 & \cos\beta \\ -\cos\alpha\sin\beta & \sin\alpha \\ -\sin\alpha\sin\beta & -\cos\alpha \\ \cos\beta & 0 \end{bmatrix}$$

The parameter ρ_a and the thermal noise power are scaled to satisfy the required SINR and INR.

In Figure 6.2 (a), the detection performance of the new Rao (or the new Wald test) is evaluated for $K < N$. The detection performance of the conventional multidimensional GLRT, Rao and Wald tests is not plotted due to the fact that the sample covariance matrix used in these detectors is singular. While the new Rao test can still work since the expression $\left(\boldsymbol{\Psi}_H \boldsymbol{S} \boldsymbol{\Psi}_H^H\right)$ is nonsingular under the condition $K > p$. In Figure 6.2 (b), the PD of the new Rao is over 0.8 for SINR $> 10\,\text{dB}$ while the conventional multidimensional detectors suffer a significant detection performance degradation. In Figure 6.2(c), the detection performance of all the detectors increases. The new Rao detector outperforms the conventional multidimensional detectors. The detection performance improvement of new Rao compared with multidimensional GLRT is about 3 dB for $P_d = 0.9$. In Figure 6.2 (d), the detection performance gap is reduced between the new Rao and the conventional multidimensional detectors. The new Rao detector still outperforms the conventional multidimensional detectors and the detection performance improvement is about 1 dB.

6.2 ADAPTIVE DETECTORS IN SUBSPACE INTERFERENCE PLUS HOMOGENEOUS CLUTTER

In this section, we consider the problem of detecting a distributed target in subspace interference plus homogeneous clutter. The signals reflected by the target are assumed to come from the same direction. However, the exact direction is unknown and the corresponding signal steering vector lies in a known subspace. The interference belongs to a known subspace, linearly independent of the signal subspace. Under the assumption, we derive the one-step Wald test and two-step Wald test. Numerical examples show that the proposed detectors can achieve better detection performance than the existing detectors.

6.2.1 Problem Formulation

Suppose the data are received by N sensors. A target, if present, is spread across K range bins. We can present the primary data as an $N \times K$ matrix $\boldsymbol{X}$. The problem is to determine whether $\boldsymbol{X}$ consists of noise $\boldsymbol{N}$ and interference $\boldsymbol{V}$ (hypothesis H_0) or noise $\boldsymbol{N}$, interference $\boldsymbol{V}$, and signal $\boldsymbol{T}$ (hypothesis H_1). The interference $\boldsymbol{V}$ lies in a subspace spanned by an $N \times q$ full-column-rank matrix $\boldsymbol{J}$ and can be represented as $\boldsymbol{V} = \boldsymbol{J}\boldsymbol{Q}$, where Q is an unknown $q \times K$ coordinate matrix. The echoes reflected from different range bins of the target have the same direction, so the signal $\boldsymbol{T}$ can be written as a rank-one matrix $\boldsymbol{T} = \boldsymbol{s}\boldsymbol{\alpha}^H$, where $\boldsymbol{s}$ is the signal steering vector and $\boldsymbol{\alpha}$ is the unknown amplitude vector. The signal steering vector $\boldsymbol{s}$ lies in a subspace spanned by an $N \times p$ full-column-rank matrix $\boldsymbol{H}$ and can be expressed as $\boldsymbol{s} = \boldsymbol{H}\boldsymbol{\theta}$, where $\boldsymbol{\theta}$ denotes the unknown $p \times 1$ coordinate vector. The noise $\boldsymbol{N}$ consists of IID columns, distributed as $\mathcal{CN}_N(\boldsymbol{0}_{N\times 1}, \boldsymbol{R})$. The estimation of $\boldsymbol{R}$ is based on L IID signal-free training data, denoted by an $N \times L$ matrix $\boldsymbol{X}_L$, which are typically obtained in the vicinity of the primary data. $\boldsymbol{X}_L$ only contains noise $\boldsymbol{N}_L$, where each column is distributed as $\mathcal{CN}_N(\boldsymbol{0}_{N\times 1}, \boldsymbol{R})$. In summary, the problem of direction detection in interference can be formulated as a binary hypothesis test

$$\begin{cases} H_0 : \boldsymbol{X} = \boldsymbol{J}\boldsymbol{Q} + \boldsymbol{N}, \boldsymbol{X}_L = \boldsymbol{N} \\ H_1 : \boldsymbol{X} = \boldsymbol{H}\boldsymbol{\theta}\boldsymbol{\alpha}^H + \boldsymbol{J}\boldsymbol{Q} + \boldsymbol{N}, \boldsymbol{X}_L = \boldsymbol{N}_L \end{cases}. \tag{6.59}$$

Note that in (6.59), $\boldsymbol{Q}$, θ, and α are unknown.

6.2.2 Wald Test-Based Detectors

In the following, we devise two detectors according to the criteria of the 1S-Wald test and 2S-Wald test.

Let $\boldsymbol{\Theta}$ be a parameter vector partitioned as

$$\boldsymbol{\Theta} = \left[\boldsymbol{\Theta}_r^T, \boldsymbol{\Theta}_s^T\right]^T, \tag{6.60}$$

where $\boldsymbol{\Theta}_r = \boldsymbol{\theta}$ and $\boldsymbol{\Theta}_s = \left[\boldsymbol{\alpha}^T, \mathrm{vec}^T\left(\boldsymbol{Q}\right)^T, \mathrm{vec}^T\left(\boldsymbol{R}\right)\right]^T$ are the relative and nuisance parameters, respectively. The FIM w.r.t $\boldsymbol{\Theta}$ is given by [7]

$$\boldsymbol{J}\left(\boldsymbol{\Theta}\right) = \mathrm{E}\left[\frac{\partial \ln f_1\left(\boldsymbol{X}, \boldsymbol{X}_L\right)}{\partial \boldsymbol{\Theta}^*}\frac{\partial \ln f_1\left(\boldsymbol{X}, \boldsymbol{X}_L\right)}{\partial \boldsymbol{\Theta}^T}\right], \tag{6.61}$$

where $f_1\left(\boldsymbol{X}, \boldsymbol{X}_L\right)$ is the joint PDF of $\boldsymbol{X}$ and $\boldsymbol{X}_L$ under hypothesis H_1. For convenience, the FIM $\boldsymbol{J}\left(\boldsymbol{\Theta}\right)$ is partitioned as

$$\boldsymbol{J}(\boldsymbol{\Theta})=\begin{bmatrix}\boldsymbol{J}_{\boldsymbol{\Theta}_r,\boldsymbol{\Theta}_r}(\boldsymbol{\Theta}), \boldsymbol{J}_{\boldsymbol{\Theta}_r,\boldsymbol{\Theta}_s}(\boldsymbol{\Theta}) \\ \boldsymbol{J}_{\boldsymbol{\Theta}_s,\boldsymbol{\Theta}_r}(\boldsymbol{\Theta}), \boldsymbol{J}_{\boldsymbol{\Theta}_s,\boldsymbol{\Theta}_s}(\boldsymbol{\Theta})\end{bmatrix}. \tag{6.62}$$

Then the Wald test is found to be Equation 6. 32. However, it is difficult to derive $\left[\boldsymbol{J}^{-1}\left(\hat{\theta}_1\right)\right]_{\theta_r,\theta_r}$. Hence, we modify the original Wald test to the following one

$$t_{\text{Wald}}=\left(\hat{\boldsymbol{\Theta}}_{r1}-\boldsymbol{\Theta}_{r0}\right)^H\left[\boldsymbol{J}_{\boldsymbol{\Theta}_r,\boldsymbol{\Theta}_r}\left(\hat{\boldsymbol{\Theta}}_1\right)\right]\left(\hat{\boldsymbol{\Theta}}_{r1}-\boldsymbol{\Theta}_{r0}\right), \tag{6.63}$$

where $\hat{\boldsymbol{\Theta}}_{r1}$ is the MLE of $\boldsymbol{\Theta}_r$ under hypothesis H_1, and $\boldsymbol{\Theta}_{r0}$ is the value of $\boldsymbol{\Theta}_r$ under hypothesis H_0. The motivation for Equation 6.63 is that in many cases, we have the fact that $\boldsymbol{J}_{\boldsymbol{\Theta}_r,\boldsymbol{\Theta}_s}(\boldsymbol{\Theta})$ is a null matrix or column [17, 18]. Hence, we arrive at $\left\{\left[\boldsymbol{J}^{-1}(\boldsymbol{\Theta})\right]_{\boldsymbol{\Theta}_r,\boldsymbol{\Theta}_r}\right\}^{-1}=\boldsymbol{J}_{\boldsymbol{\Theta}_r,\boldsymbol{\Theta}_r}(\boldsymbol{\Theta})$. In this case, the modified Wald test in Equation 6.63 is the same as the one in Equation 6.32.

The joint PDF $f_1(\boldsymbol{X},\boldsymbol{X}_L)$ is given by

$$f_1(\boldsymbol{X},\boldsymbol{X}_L)=\left(\pi^N|\boldsymbol{R}|\right)^{-(K+L)}\exp\left\{-\text{tr}(\boldsymbol{R}^{-1}\boldsymbol{S})-\text{tr}\left[(\boldsymbol{X}-\boldsymbol{DE})^H\boldsymbol{R}^{-1}(\boldsymbol{X}-\boldsymbol{DE})\right]\right\}, \tag{6.64}$$

where $\boldsymbol{S}=\boldsymbol{X}_L\boldsymbol{X}_L^H$ is L times the sample covariance matrix,

$$\boldsymbol{D}=[\boldsymbol{s},\boldsymbol{J}], \tag{6.65}$$

and

$$\boldsymbol{E}=\begin{bmatrix}\boldsymbol{\alpha}^H \\ \boldsymbol{Q}\end{bmatrix}. \tag{6.66}$$

Taking the logarithm of Equation 6.64 and performing the derivative w.r.t. $\boldsymbol{\theta}^*$ and $\boldsymbol{\theta}^T$ yields

$$\frac{\partial \ln f_1(\boldsymbol{X},\boldsymbol{X}_L)}{\partial\boldsymbol{\theta}^*}=\boldsymbol{H}^H\boldsymbol{R}^{-1}\boldsymbol{X}_1\boldsymbol{\alpha} \tag{6.67}$$

and

$$\frac{\partial \ln f_1\left(\boldsymbol{X},\boldsymbol{X}_L\right)}{\partial \theta^T} = \boldsymbol{\alpha}\boldsymbol{X}_1^H\boldsymbol{R}^{-1}\boldsymbol{H} \tag{6.68}$$

respectively, where

$$\boldsymbol{X}_1 = \boldsymbol{X} - \boldsymbol{D}\boldsymbol{E} \tag{6.69}$$

Substituting (6.67) and (6.68) into (6.61) results in

$$\boldsymbol{J}_{\Theta_r,\Theta_r}\left(\Theta\right) = \boldsymbol{H}^H\boldsymbol{R}^{-1}\mathrm{E}\left[\boldsymbol{X}_1\boldsymbol{\alpha}\boldsymbol{\alpha}^H\boldsymbol{X}_1^H\right]\boldsymbol{R}^{-1}\boldsymbol{H}. \tag{6.70}$$

One can verify that

$$\mathrm{E}\left[\boldsymbol{X}_1\boldsymbol{\alpha}\boldsymbol{\alpha}^H\boldsymbol{X}_1^H\right] = \boldsymbol{\alpha}^H\boldsymbol{\alpha}\cdot\boldsymbol{R} \tag{6.71}$$

Plugging (6.71) to (6.70) results in

$$\boldsymbol{J}_{\Theta_r,\Theta_r}\left(\Theta\right) = \boldsymbol{\alpha}^H\boldsymbol{\alpha}\cdot\boldsymbol{H}^H\boldsymbol{R}^{-1}\boldsymbol{H}. \tag{6.72}$$

Substituting (6.72) into (6.63), along with the fact $\Theta_{r0} = \mathbf{0}_{p\times 1}$, leads to the Wald test, for fixed θ, α, and $\boldsymbol{R}$ as

$$t_{\text{Wald}|\theta,\alpha,\boldsymbol{R}} = \boldsymbol{\alpha}^H\boldsymbol{\alpha}\cdot\theta^H\boldsymbol{H}^H\boldsymbol{R}^{-1}\boldsymbol{H}\theta, \tag{6.73}$$

where we have added the subscript "$|\theta,\alpha,\boldsymbol{R}$" to indicate the dependence of the Wald test on θ, α, and $\boldsymbol{R}$.

6.2.2.1 One-Step Wald test

In order to obtain the explicit Wald test, we need to derive the MLEs of θ, α, and $\boldsymbol{R}$ under hypothesis H_1. The MLE of $\boldsymbol{R}$ in Equation 6.64, with θ and α fixed, is well known and is given by

$$\hat{\boldsymbol{R}}_1 = \frac{1}{K+L}\left(\boldsymbol{S} + \boldsymbol{X}_1\boldsymbol{X}_1^H\right). \tag{6.74}$$

Plugging (6.74) into (6.64) yields

$$f_1\left(\boldsymbol{X},\boldsymbol{X}_L;\hat{\boldsymbol{R}}_1\right)=c\left|\boldsymbol{S}+\left(\boldsymbol{X}-\boldsymbol{D}\boldsymbol{E}\right)\left(\boldsymbol{X}-\boldsymbol{D}\boldsymbol{E}\right)^H\right|^{-(K+L)}, \tag{6.75}$$

where $c=\left[(K+L)/(e\pi)\right]^{N(K+L)}$. Exploiting the identity [8]

$$\left|\boldsymbol{A}+\boldsymbol{B}\boldsymbol{C}\boldsymbol{L}\right|=\left|\boldsymbol{A}\right|\cdot\left|\boldsymbol{C}\right|\cdot\left|\boldsymbol{L}\boldsymbol{A}^{-1}\boldsymbol{B}+\boldsymbol{C}^{-1}\right| \tag{6.76}$$

for any conformable matrices involved, we can rewrite Equation 6.75 as

$$f_1\left(\boldsymbol{X},\boldsymbol{X}_L;\theta_1\right)=c_S\left|\boldsymbol{I}_K+\left(\tilde{\boldsymbol{X}}-\tilde{\boldsymbol{D}}\boldsymbol{E}\right)^H\left(\tilde{\boldsymbol{X}}-\tilde{\boldsymbol{D}}\boldsymbol{E}\right)\right|^{-(K+L)}, \tag{6.77}$$

where $c_S=c\left|\boldsymbol{S}\right|^{-(K+L)}$, $\tilde{\boldsymbol{X}}=\boldsymbol{S}^{-1/2}\boldsymbol{X}$, and $\tilde{\boldsymbol{D}}=\boldsymbol{S}^{-1/2}\boldsymbol{D}$.

There is an ambiguity for the MLEs of $\boldsymbol{\theta}$ and $\boldsymbol{\alpha}$ in Equation 6.77. To eliminate this ambiguity, we put the following constraint on $\boldsymbol{\theta}$

$$\boldsymbol{\theta}^H\boldsymbol{H}^H\boldsymbol{H}\boldsymbol{\theta}=1. \tag{6.78}$$

Nulling the derivative of Equation 6.77 w.r.t. $\boldsymbol{E}$ yields the MLE of $\boldsymbol{E}$

$$\hat{\boldsymbol{E}}_1=\left(\tilde{\boldsymbol{D}}^H\boldsymbol{D}\right)^{-1}\tilde{\boldsymbol{D}}^H\tilde{\boldsymbol{X}}. \tag{6.79}$$

Plugging Equation 6.79 into 6.74, after some algebra, leads to

$$\begin{aligned}\hat{\boldsymbol{R}}_1&=\frac{1}{K+L}\boldsymbol{S}^{1/2}\left(\boldsymbol{I}_N+\tilde{\boldsymbol{X}}\tilde{\boldsymbol{X}}^H-\tilde{\boldsymbol{X}}\tilde{\boldsymbol{X}}^H\boldsymbol{P}_{\tilde{\boldsymbol{D}}}-\boldsymbol{P}_{\tilde{\boldsymbol{D}}}\tilde{\boldsymbol{X}}\tilde{\boldsymbol{X}}^H+\boldsymbol{P}_{\tilde{\boldsymbol{D}}}\tilde{\boldsymbol{X}}\tilde{\boldsymbol{X}}^H\boldsymbol{P}_{\tilde{\boldsymbol{D}}}\right)\boldsymbol{S}^{1/2}\\&\frac{1}{K+L}\boldsymbol{S}^{1/2}\left(\boldsymbol{I}_N+\boldsymbol{P}_{\tilde{\boldsymbol{D}}}^{\perp}\tilde{\boldsymbol{X}}\tilde{\boldsymbol{X}}^H\boldsymbol{P}_{\tilde{\boldsymbol{D}}}^{\perp}\right)\boldsymbol{S}^{1/2}\end{aligned}, \tag{6.80}$$

where

$$\boldsymbol{P}_{\tilde{\boldsymbol{D}}}=\tilde{\boldsymbol{D}}\left(\tilde{\boldsymbol{D}}^H\tilde{\boldsymbol{D}}\right)^{-1}\tilde{\boldsymbol{D}}^H \tag{6.81}$$

is the orthogonal projection matrix of $\tilde{\boldsymbol{D}}$ and $\boldsymbol{P}_{\tilde{\boldsymbol{D}}}^{\perp}=\boldsymbol{I}_N-\boldsymbol{P}_{\tilde{\boldsymbol{D}}}$. Performing the matrix inversion operation to Equation 6.80, we have

$$\hat{\boldsymbol{R}}_1^{-1}=(K+L)\boldsymbol{S}^{-1/2}\left[\boldsymbol{I}_N-\boldsymbol{P}_{\tilde{\boldsymbol{D}}}^{\perp}\tilde{\boldsymbol{X}}\left(\boldsymbol{I}_K+\tilde{\boldsymbol{X}}^H\boldsymbol{P}_{\tilde{\boldsymbol{D}}}^{\perp}\tilde{\boldsymbol{X}}\right)^{-1}\tilde{\boldsymbol{X}}^H\boldsymbol{P}_{\tilde{\boldsymbol{D}}}^{\perp}\right]\boldsymbol{S}^{-1/2} \tag{6.82}$$

It follows from (6.82) that

$$\boldsymbol{\theta}^H\boldsymbol{H}^H\hat{\boldsymbol{R}}_1^{-1}\boldsymbol{H}\boldsymbol{\theta}=(K+L)\left[\boldsymbol{\theta}^H\tilde{\boldsymbol{H}}^H\tilde{\boldsymbol{H}}\boldsymbol{\theta}-\boldsymbol{\theta}^H\tilde{\boldsymbol{H}}\boldsymbol{P}_{\tilde{\boldsymbol{D}}}^{\perp}\tilde{\boldsymbol{X}}\left(\boldsymbol{I}_K+\tilde{\boldsymbol{X}}^H\boldsymbol{P}_{\tilde{\boldsymbol{D}}}^{\perp}\tilde{\boldsymbol{X}}\right)^{-1}\tilde{\boldsymbol{X}}^H\boldsymbol{P}_{\tilde{\boldsymbol{D}}}^{\perp}\tilde{\boldsymbol{H}}\boldsymbol{\theta}\right]. \tag{6.83}$$

According to the definition of $\boldsymbol{D}$ in (6.65), we have [19]

$$\boldsymbol{P}_{\tilde{\boldsymbol{D}}}=\boldsymbol{P}_{\tilde{\boldsymbol{J}}}+\boldsymbol{P}_{\boldsymbol{P}_{\tilde{\boldsymbol{J}}}^{\perp}\tilde{\boldsymbol{s}}}. \tag{6.84}$$

It follows that

$$\boldsymbol{P}_{\tilde{\boldsymbol{D}}}^{\perp}=\boldsymbol{P}_{\tilde{\boldsymbol{J}}}^{\perp}-\boldsymbol{P}_{\mathbf{P}_{\tilde{\boldsymbol{J}}}^{\perp}\tilde{\boldsymbol{s}}}. \tag{6.85}$$

Post-multiplying (6.85) by $\tilde{\boldsymbol{s}}$, we have

$$\boldsymbol{P}_{\tilde{\boldsymbol{D}}}^{\perp}\tilde{\boldsymbol{s}}=\boldsymbol{P}_{\tilde{\boldsymbol{J}}}^{\perp}\tilde{\boldsymbol{s}}-\boldsymbol{P}_{\tilde{\boldsymbol{J}}}^{\perp}\tilde{\boldsymbol{s}}\tilde{\boldsymbol{s}}^H\boldsymbol{P}_{\tilde{\boldsymbol{J}}}^{\perp}\tilde{\boldsymbol{s}}\,/\,\tilde{\boldsymbol{s}}^H\boldsymbol{P}_{\tilde{\boldsymbol{J}}}^{\perp}\tilde{\boldsymbol{s}}=\mathbf{0}_{N\times1}. \tag{6.86}$$

Plugging (6.86) into (6.83), along with $\tilde{\boldsymbol{s}}=\tilde{\boldsymbol{H}}\boldsymbol{\theta}$, yields

$$\boldsymbol{\theta}^H\boldsymbol{H}^H\tilde{\boldsymbol{R}}_1^{-1}\boldsymbol{H}\boldsymbol{\theta}=(K+L)\boldsymbol{\theta}^H\tilde{\boldsymbol{H}}^H\tilde{\boldsymbol{H}}\boldsymbol{\theta}. \tag{6.87}$$

Substituting (6.87) into (6.73) leads to

$$t_{\text{Wald}|\boldsymbol{\theta},\boldsymbol{\alpha}}=\boldsymbol{\alpha}^H\boldsymbol{\alpha}\cdot\boldsymbol{\theta}^H\tilde{\boldsymbol{H}}^H\tilde{\boldsymbol{H}}\boldsymbol{\theta} \tag{6.88}$$

where we have dropped the constant.

Note that the MLE of $\boldsymbol{\alpha}^H$ corresponds to the first row of Equation 6.79. Using the partitioned matrix inversion formula [20], we have

$$\hat{\alpha}_1^H=\frac{\tilde{\boldsymbol{s}}^H\boldsymbol{P}_{\tilde{\boldsymbol{J}}}^{\perp}\tilde{\boldsymbol{X}}}{\tilde{\boldsymbol{s}}^H\boldsymbol{P}_{\tilde{\boldsymbol{J}}}^{\perp}\tilde{\boldsymbol{s}}} \tag{6.89}$$

Substituting Equation 6.89 into Equation 6.88 results in

$$t_{\text{Wald}|\,\boldsymbol{\theta}}=\frac{\tilde{\boldsymbol{s}}^H\tilde{\boldsymbol{s}}\cdot\tilde{\boldsymbol{s}}^H\boldsymbol{P}_{\tilde{\boldsymbol{J}}}^{\perp}\tilde{\boldsymbol{X}}\tilde{\boldsymbol{X}}^H\boldsymbol{P}_{\tilde{\boldsymbol{J}}}^{\perp}\tilde{\boldsymbol{s}}}{\left(\tilde{\boldsymbol{s}}^H\boldsymbol{P}_{\tilde{\boldsymbol{J}}}^{\perp}\tilde{\boldsymbol{s}}\right)^2}, \tag{6.90}$$

where we have used $\tilde{\boldsymbol{s}} = \tilde{\boldsymbol{H}}\boldsymbol{\theta}$.

Now we proceed to derive the MLE of θ under hypothesis H_1. Inserting (6.79) into (6.77), after some algebra, yields

$$f_1\left(\boldsymbol{X},\boldsymbol{X}_L;\boldsymbol{R}_1,\boldsymbol{E}_1\right) = c_S\left|\boldsymbol{I}_K + \tilde{\boldsymbol{X}}^H \boldsymbol{P}_{\tilde{\boldsymbol{D}}}^{\perp}\tilde{\boldsymbol{X}}\right|^{-(K+L)}. \tag{6.91}$$

According to Equation 6.85, we can rewrite Equation 6.91 as

$$f_1\left(\boldsymbol{X},\boldsymbol{X}_L;\boldsymbol{R}_1,\boldsymbol{E}_1\right) = c_S\left|\boldsymbol{I}_K + \tilde{\boldsymbol{X}}^H \boldsymbol{P}_{\tilde{\boldsymbol{J}}}^{\perp}\tilde{\boldsymbol{X}} - \tilde{\boldsymbol{X}}^H \boldsymbol{P}_{\boldsymbol{P}_{\tilde{\boldsymbol{J}}}^{\perp}\tilde{\boldsymbol{s}}}\tilde{\boldsymbol{X}}\right|^{-(K+L)}. \tag{6.92}$$

Moreover, we observe that

$$\left|\boldsymbol{I}_K + \tilde{\boldsymbol{X}}^H \boldsymbol{P}_{\tilde{\boldsymbol{J}}}^{\perp}\tilde{\boldsymbol{X}} - \tilde{\boldsymbol{X}}^H \boldsymbol{P}_{\boldsymbol{P}_{\tilde{\boldsymbol{J}}}^{\perp}\tilde{\boldsymbol{s}}}\tilde{\boldsymbol{X}}\right| = \left|\boldsymbol{I}_K + \tilde{\boldsymbol{X}}^H \boldsymbol{P}_{\tilde{\boldsymbol{J}}}^{\perp}\tilde{\boldsymbol{X}} - \frac{\tilde{\boldsymbol{X}}^H \boldsymbol{P}_{\tilde{\boldsymbol{J}}}^{\perp}\tilde{\boldsymbol{s}}^H \boldsymbol{P}_{\tilde{\boldsymbol{J}}}^{\perp}\tilde{\boldsymbol{X}}}{\tilde{\boldsymbol{s}}^H \boldsymbol{P}_{\tilde{\boldsymbol{J}}}^{\perp}\tilde{\boldsymbol{s}}}\right|. \tag{6.93}$$

Using (6.76), we can rewrite Equation 6.93 as

$$\begin{aligned}&\left|\boldsymbol{I}_K + \tilde{\boldsymbol{X}}^H \boldsymbol{P}_{\tilde{\boldsymbol{J}}}^{\perp}\tilde{\boldsymbol{X}} - \tilde{\boldsymbol{X}}^H \boldsymbol{P}_{\boldsymbol{P}_{\tilde{\boldsymbol{J}}}^{\perp}}^{\perp}\tilde{\boldsymbol{X}}\right| \\ &= \left|\boldsymbol{I}_K + \tilde{\boldsymbol{X}}^H \boldsymbol{P}_{\tilde{\boldsymbol{J}}}^{\perp}\tilde{\boldsymbol{X}}\right| \cdot \left[1 - \frac{\tilde{\boldsymbol{s}}^H \boldsymbol{P}_{\tilde{\boldsymbol{J}}}^{\perp}\tilde{\boldsymbol{X}}\left(\boldsymbol{I}_K + \tilde{\boldsymbol{X}}^H \boldsymbol{P}_{\tilde{\boldsymbol{J}}}^{\perp}\tilde{\boldsymbol{X}}\right)^{-1}\tilde{\boldsymbol{X}}^H \boldsymbol{P}_{\tilde{\boldsymbol{J}}}^{\perp}\tilde{\boldsymbol{s}}}{\tilde{\boldsymbol{s}}^H \boldsymbol{P}_{\tilde{\boldsymbol{J}}}^{\perp}\tilde{\boldsymbol{s}}}\right],\end{aligned} \tag{6.94}$$

which, with $\boldsymbol{s}$ being substituted by $\boldsymbol{H}\boldsymbol{\theta}$, becomes

$$\begin{aligned}&\left|\boldsymbol{I}_K + \tilde{\boldsymbol{X}}^H \boldsymbol{P}_{\tilde{\boldsymbol{J}}}^{\perp}\tilde{\boldsymbol{X}} - \tilde{\boldsymbol{X}}^H \boldsymbol{P}_{\boldsymbol{P}_{\tilde{\boldsymbol{J}}}^{\perp}}^{\perp}\tilde{\boldsymbol{X}}\right| \\ &= \left|\boldsymbol{I}_K + \tilde{\boldsymbol{X}}^H \boldsymbol{P}_{\tilde{\boldsymbol{J}}}^{\perp}\tilde{\boldsymbol{X}}\right| \cdot \left[1 - \frac{\boldsymbol{\theta}^H \tilde{\boldsymbol{H}}^H \boldsymbol{P}_{\tilde{\boldsymbol{J}}}^{\perp}\tilde{\boldsymbol{X}}\left(\boldsymbol{I}_K + \tilde{\boldsymbol{X}}^H \boldsymbol{P}_{\tilde{\boldsymbol{J}}}^{\perp}\tilde{\boldsymbol{X}}\right)^{-1}\tilde{\boldsymbol{X}}^H \boldsymbol{P}_{\tilde{\boldsymbol{J}}}^{\perp}\tilde{\boldsymbol{H}}\boldsymbol{\theta}}{\boldsymbol{\theta}^H \tilde{\boldsymbol{H}}^H \boldsymbol{P}_{\tilde{\boldsymbol{J}}}^{\perp}\tilde{\boldsymbol{H}}\boldsymbol{\theta}}\right].\end{aligned} \tag{6.95}$$

The maximum value of (6.95) w.r.t. θ is

$$\left|\boldsymbol{I}_K + \tilde{\boldsymbol{X}}^H \boldsymbol{P}_{\tilde{\boldsymbol{J}}}^{\perp}\tilde{\boldsymbol{X}} - \tilde{\boldsymbol{X}}^H \boldsymbol{P}_{\boldsymbol{P}_{\tilde{\boldsymbol{J}}}^{\perp}}^{\perp}\tilde{\boldsymbol{X}}\right| = \left|\boldsymbol{I}_K + \tilde{\boldsymbol{X}}^H \boldsymbol{P}_{\tilde{\boldsymbol{J}}}^{\perp}\tilde{\boldsymbol{X}}\right| \cdot \left[1 - \lambda_{\max}\left(\boldsymbol{F}\right)\right], \tag{6.96}$$

where $\lambda_{\max}(\boldsymbol{F})$ denotes the maximum eigenvalue of $\boldsymbol{F}$, defined by

$$\boldsymbol{F} = \tilde{\boldsymbol{H}}^H \boldsymbol{P}_{\tilde{J}}^{\perp} \tilde{\boldsymbol{X}} \left(\boldsymbol{I}_K + \tilde{\boldsymbol{X}}^H \boldsymbol{P}_{\tilde{J}}^{\perp} \tilde{\boldsymbol{X}} \right)^{-1} \tilde{\boldsymbol{X}}^H \boldsymbol{P}_{\tilde{J}}^{\perp} \tilde{\boldsymbol{H}} \left(\tilde{\boldsymbol{H}}^H \boldsymbol{P}_{\tilde{J}}^{\perp} \tilde{\boldsymbol{H}} \right)^{-1}. \tag{6.97}$$

Moreover, the MLE of $\boldsymbol{\theta}$ under the constraint of (6.78) is

$$\hat{\boldsymbol{\theta}} = \frac{\boldsymbol{\theta}_F}{\left(\boldsymbol{\theta}_F^H \tilde{\boldsymbol{H}}^H \tilde{\boldsymbol{H}} \boldsymbol{\theta}_F \right)^{1/2}}, \tag{6.98}$$

where $\boldsymbol{\theta}_F$ is a principal eigenvector of the matrix $\boldsymbol{F}$ in (6.97) with an arbitrary norm. Therefore, plugging (6.98) into (6.90) results in the final Wald test

$$t_{\text{Wald}} = \frac{\boldsymbol{\theta}_F^H \tilde{\boldsymbol{H}}^H \boldsymbol{P}_{\tilde{J}}^{\perp} \tilde{\boldsymbol{X}} \tilde{\boldsymbol{X}}^H \boldsymbol{P}_{\tilde{J}}^{\perp} \tilde{\boldsymbol{H}} \boldsymbol{\theta}_F \cdot \boldsymbol{\theta}_F^H \tilde{\boldsymbol{H}}^H \tilde{\boldsymbol{H}} \boldsymbol{\theta}_F}{\boldsymbol{\theta}_F^H \tilde{\boldsymbol{H}}^H \boldsymbol{P}_{\tilde{J}}^{\perp} \tilde{\boldsymbol{H}} \boldsymbol{\theta}_F}. \tag{6.99}$$

Note that if we define

$$\tilde{\boldsymbol{s}}_F = \tilde{\boldsymbol{H}} \boldsymbol{\theta}_F, \tag{6.100}$$

then (6.99) can be rewritten as

$$t_{\text{Wald}} = \text{tr}\left(\tilde{\boldsymbol{X}}^H \boldsymbol{P}_{\tilde{s}_F | \tilde{J}}^H \boldsymbol{P}_{\tilde{s}_F | \tilde{J}} \tilde{\boldsymbol{X}} \right), \tag{6.101}$$

where

$$\boldsymbol{P}_{\tilde{s}_F | \tilde{J}} = \frac{\tilde{\boldsymbol{s}}_F \tilde{\boldsymbol{s}}_F^H \boldsymbol{P}_{\tilde{J}}^{\perp}}{\tilde{\boldsymbol{s}}_F^H \boldsymbol{P}_{\tilde{J}}^{\perp} \tilde{\boldsymbol{s}}_F} \tag{6.102}$$

is the oblique projection matrix onto the subspace $\langle \tilde{\boldsymbol{s}}_F \rangle$ along the subspace by $\langle \tilde{\boldsymbol{J}} \rangle$.

6.2.2.2 Two-Step Wald Test

Nulling the derivative of Equation 6.64 w.r.t. $\boldsymbol{E}$ yields the MLE of $\boldsymbol{E}$ for given $\boldsymbol{R}$ under hypothesis H_1 as

$$\hat{\boldsymbol{E}}_1 = \left(\bar{\boldsymbol{D}}^H \bar{\boldsymbol{D}} \right)^{-1} \bar{\boldsymbol{D}}^H \bar{\boldsymbol{X}} \tag{6.103}$$

where $\bar{D} = R^{-1/2}D$ and $\bar{X} = R^{-1/2}X$. The MLE of α for given R is the first column of $\hat{E}_1$ in (6.103). In a manner similar to (6.89), we have

$$\hat{\alpha}_1^H = \frac{\bar{s}^H P_{\bar{J}}^{\perp} \bar{X}}{\bar{s}^H P_{\bar{J}}^{\perp} \bar{s}}, \tag{6.104}$$

where $\bar{s} = R^{-1/2}s, \bar{J} = R^{-1/2}J, P_{\bar{J}} = \bar{J}\left(\bar{J}^H\bar{J}\right)^{-1}\bar{J}^H$, and $P_{\bar{J}}^{\perp} = I_N - P_{\bar{J}}$. Substituting Equation 6.104 into Equation 6.73 leads to

$$t_{\text{Wald}|\,\theta,R} = \frac{\bar{s}^H P_{\bar{J}}^{\perp} \bar{X}\bar{X}^H P_{\bar{J}}^{\perp} \bar{s} \cdot \theta^H \bar{H}^H \bar{H}\theta}{\left(\bar{s}^H P_{\bar{J}}^{\perp} \bar{s}\right)^2}. \tag{6.105}$$

We proceed to derive the MLE of θ for given R. It results from (6.64) and (6.103) that

$$f_1\left(X; E_1\right) = \left(\pi^N |R|\right)^{-K} \exp\left[-\text{tr}\left(\bar{X}^H P_{\bar{D}}^{\perp} \bar{X}\right)\right], \tag{6.106}$$

where $P_{\bar{D}} = \bar{D}\left(\bar{D}^H\bar{D}\right)^{-1}\bar{D}^H$ and $P_{\bar{D}}^{\perp} = I_N - P_{\bar{D}}$. Analogous to Equation 6.85, we have

$$P_{\bar{D}}^{\perp} = P_{\bar{J}}^{\perp} - P_{P_{\bar{J}}^{\perp}\bar{s}}. \tag{6.107}$$

Moreover, according to $\bar{s} = \bar{H}\theta$, Equation 6.106 can be recast as

$$f_1\left(X; \hat{E}_1\right) = \left(\pi^N |R|\right)^{-K} \exp\left[-\text{tr}\left(\bar{X}^H P_{\bar{J}}^{\perp} \bar{X}\right)\right] \cdot \exp\left(\frac{\theta^H \bar{H}^H P_{\bar{J}}^{\perp} \bar{X}\bar{X}^H P_{\bar{J}}^{\perp} \bar{H}\theta}{\theta^H \bar{H}^H P_{\bar{J}}^{\perp} \bar{H}\theta}\right), \tag{6.108}$$

which has a similar form as Equation 6.95. Hence, we can analogously have the MLE of θ in Equation 6.108 for given R under the constraint Equation 6.78, which is given by

$$\hat{\theta}_R = \frac{\theta_K}{\left(\theta_R^H H_R^H H \theta_{K_R}\right)^{1/2}}, \tag{6.109}$$

where θ_{K_R} is a principal eigenvector of the matrix

$$\boldsymbol{K}_{\boldsymbol{R}} = \bar{\boldsymbol{H}}^H \boldsymbol{P}_{\bar{J}}^{\perp} \bar{\boldsymbol{X}} \bar{\boldsymbol{X}}^H \boldsymbol{P}_{\bar{J}}^{\perp} \bar{\boldsymbol{H}} \left(\bar{\boldsymbol{H}}^H \boldsymbol{P}_{\bar{J}}^{\perp} \bar{\boldsymbol{H}} \right)^{-1}. \tag{6.110}$$

Plugging (6.109) into (6.105), along with $\boldsymbol{s} = \boldsymbol{H}\boldsymbol{\theta}$, results in the Wald test for given $\boldsymbol{R}$

$$t_{\text{Wald}|\boldsymbol{R}} = \frac{\boldsymbol{\theta}_{K_R}^H \bar{\boldsymbol{H}}^H \boldsymbol{P}_{\bar{J}}^{\perp} \bar{\boldsymbol{X}} \bar{\boldsymbol{X}}^H \boldsymbol{P}_{\bar{J}}^{\perp} \bar{\boldsymbol{H}} \boldsymbol{\theta}_{K_R} \cdot \boldsymbol{\theta}_{K}^H \bar{\boldsymbol{H}}^H \bar{\boldsymbol{H}} \boldsymbol{\theta}_{K}^H}{\left(\boldsymbol{\theta}_{K_R}^H \bar{\boldsymbol{H}}^H \boldsymbol{P}_{\bar{J}}^{\perp} \bar{\boldsymbol{H}} \boldsymbol{\theta}_{K_R} \right)^2}. \tag{6.111}$$

It is well known that the MLE of $\boldsymbol{R}$, based on the training data is $\hat{\boldsymbol{R}} = \boldsymbol{S} / L$, where $\mathbf{S}$ is given in (6.64). Substituting $\hat{\boldsymbol{R}}$ for the implicitly used $\boldsymbol{R}$ in Equation 6.111 and dropping the constant leads to the final 2S-Wald test

$$t_{\text{2S-Wald}} = \frac{\boldsymbol{\theta}_{K}^H \tilde{\boldsymbol{H}}^H \boldsymbol{P}_{\tilde{J}}^{\perp} \tilde{\boldsymbol{X}} \tilde{\boldsymbol{X}}^H \boldsymbol{P}_{\tilde{J}}^{\perp} \tilde{\boldsymbol{H}} \boldsymbol{\theta}_{K} \cdot \boldsymbol{\theta}_{K}^H \tilde{\boldsymbol{H}}^H \tilde{\boldsymbol{H}} \boldsymbol{\theta}_{K}^H}{\left(\boldsymbol{\theta}_{K}^H \tilde{\boldsymbol{H}}^H \boldsymbol{P}_{\tilde{J}}^{\perp} \tilde{\boldsymbol{H}} \boldsymbol{\theta}_{K} \right)^2}, \tag{6.112}$$

where θ_K is a principal vector of

$$\boldsymbol{K} = \tilde{\boldsymbol{H}}^H \boldsymbol{P}_{\tilde{J}}^{\perp} \tilde{\boldsymbol{X}} \tilde{\boldsymbol{X}}^H \boldsymbol{P}_{\tilde{J}}^{\perp} \tilde{\boldsymbol{H}} \left(\tilde{\boldsymbol{H}}^H \boldsymbol{P}_{\tilde{J}}^{\perp} \tilde{\boldsymbol{H}} \right)^{-1} \tag{6.113}$$

which is a modified version of $\boldsymbol{K}_{\boldsymbol{R}}$ in (6.110), with $\boldsymbol{R}$ being replaced by $\boldsymbol{S}$.

Equation (6.112) can be rewritten as

$$t_{\text{2S-Wald}} = \operatorname{tr}\left(\tilde{\boldsymbol{X}}^H \boldsymbol{P}_{\tilde{\boldsymbol{s}}_K | \tilde{J}}^H \boldsymbol{P}_{\tilde{\boldsymbol{s}}_K | \tilde{J}} \tilde{\boldsymbol{X}} \right), \tag{6.114}$$

where

$$\tilde{\boldsymbol{s}}_K = \tilde{\boldsymbol{H}} \boldsymbol{\theta}_K, \tag{6.115}$$

and

$$\boldsymbol{P}_{\tilde{\boldsymbol{s}}_K | \tilde{J}} = \frac{\tilde{\boldsymbol{s}}_K \tilde{\boldsymbol{s}}_K^H \boldsymbol{P}_{\tilde{J}}^{\perp}}{\tilde{\boldsymbol{s}}_K^H \boldsymbol{P}_{\tilde{J}}^{\perp} \tilde{\boldsymbol{s}}_K} \tag{6.116}$$

is the oblique projection matrix onto the subspace $\langle \tilde{\boldsymbol{s}}_K \rangle$ along the subspace $\langle \tilde{\boldsymbol{J}} \rangle$.

6.2.3 Numerical Examples

In this section, we evaluate the detection performance of the proposed detectors by Monte Carlo simulations. For comparison purposes, we also consider the GLRT and 2S-GLRT for the detection problem in Equation 6.59, which are found to be [21][1]

$$t_{\text{GLRT}} = \lambda_{\max}\left[\boldsymbol{P}_{\boldsymbol{Z}^H\tilde{\boldsymbol{H}}}\boldsymbol{Z}^H\tilde{\boldsymbol{X}}\left(\boldsymbol{I}_K + \tilde{\boldsymbol{X}}^H\boldsymbol{P}_{\tilde{\boldsymbol{J}}}^{\perp}\tilde{\boldsymbol{X}}\right)^{-1}\tilde{\boldsymbol{X}}^H\boldsymbol{Z}\boldsymbol{P}_{\boldsymbol{Z}^H\tilde{\boldsymbol{H}}}\right] \tag{6.117}$$

and

$$t_{\text{2S-GLRT}} = \lambda_{\max}\left(\boldsymbol{P}_{\boldsymbol{Z}^H\tilde{\boldsymbol{H}}}\boldsymbol{Z}^H\tilde{\boldsymbol{X}}\tilde{\boldsymbol{X}}^H\boldsymbol{Z}\boldsymbol{P}_{\boldsymbol{Z}^H\tilde{\boldsymbol{H}}}\right), \tag{6.118}$$

respectively, where $\boldsymbol{Z}$ is a semi-unitary matrix such as $\boldsymbol{Z}\boldsymbol{Z}^H = \boldsymbol{P}_{\tilde{\boldsymbol{J}}}^{\perp}$ and $\boldsymbol{Z}^H\boldsymbol{Z} = \boldsymbol{I}_{N-q}$.

In all figures, the quantities $\boldsymbol{H}, \boldsymbol{\theta}, \boldsymbol{\alpha}, \boldsymbol{J}$, and $\boldsymbol{Q}$ are randomly chosen. However, they never change after they are generated. We set $N = 8$ and PFA$=10^{-3}$. To evaluate the PD and detection threshold to ensure a preassigned PFA, we run 10^4 and 10^6 trails, respectively. The (i, j)th entry of the covariance matrix $\boldsymbol{R}$ is set to be

$$\boldsymbol{R}_{i,j} = \sigma_n^2 \varepsilon^{\text{abs}(i-j)}, i, j = 1, 2, \ldots, N, \tag{6.119}$$

where σ_n^2 and ε are two positive scalars. Hence, σ_n^2 and ε can be taken as the power parameter and structure parameter of noise, respectively.

Figure 6.3 compares the PDs of the detectors under different SNRs, defined as

$$\text{SNR} = \boldsymbol{\alpha}^H\boldsymbol{\alpha} \cdot \boldsymbol{\theta}^H\boldsymbol{H}^H\boldsymbol{R}^{-1}\boldsymbol{H}\boldsymbol{\theta}. \tag{6.120}$$

Moreover, the INR is similarly defined as

$$\text{INR} = \boldsymbol{\phi}^H\boldsymbol{J}^H\boldsymbol{R}^{-1}\boldsymbol{J}\boldsymbol{\phi}. \tag{6.121}$$

The results in Figure 6.3 indicate that the 1S-Wald and 2S-Wald tests roughly achieve the same detection performance. However, the PDs of the

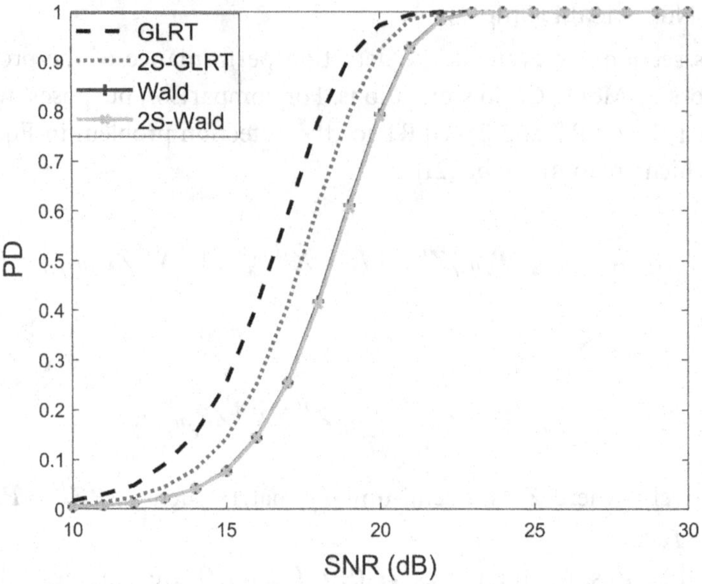

FIGURE 6.3 PDs of the detectors under different SNRs. $p=2$, $q=1$, $K=6$, $L=16$, $\sigma_n^2=1$, $\varepsilon=0.95$, and INR = 15dB.

1S-Wald and 2S-Wald are lower than that of the 2S-GLRT, which, in turn, is lower than the PD of the 1S-GLRT.

In Figure 6.4, the number of training data increases. The results indicate that the 1S-Wald and 2S-Wald tests also roughly achieve the same detection performance and they have slightly better detection performance than the 1S-GLRT, which, in turn, provides slightly better detection performance than the 2S-GLRT. Comparing the results in Figures 6.3 and 6.4 indicates that the PDs of all the detectors increase as the number of training data increases.

In Figure 6.5, the dimension of the interference subspace increases. The results show that the 1S-Wald and 2S-Wald tests can provide slightly better detection performance than the 1S-GLRT and 2S-GLRT. Moreover, the PDs of all the detectors increase as the dimension of the interference subspace increases. This is owing to the increase in the loss of the signal energy projected onto the interference subspace.

Figure 6.6 displays the PDs of the detectors under different INRs. It can be seen that the PDs of the detectors are nearly unchanged as the INR increases. In other words, all the detectors can effectively reject the interference. In addition, for the chosen parameters, the 1S-Wald and 2S-Wald

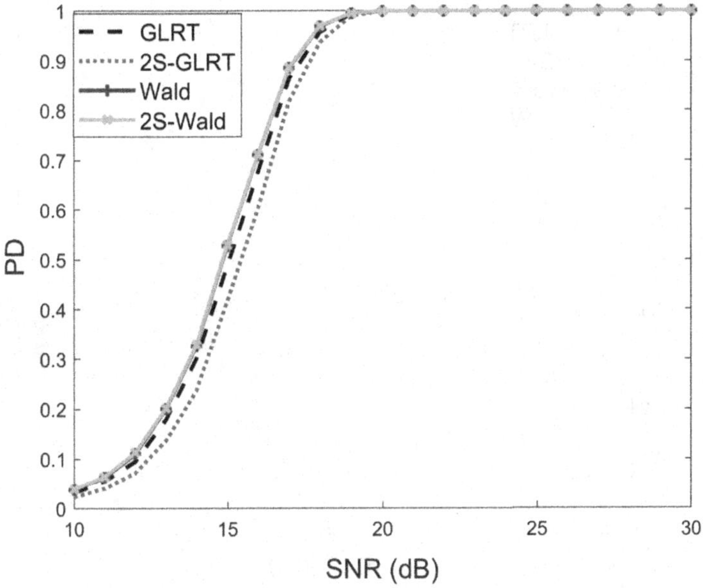

FIGURE 6.4 PDs of the detectors under different SNRs. $p=2, q=1, K=6, L=32$, $\sigma_n^2=1$, $\varepsilon=0.95$, and $\mathrm{INR}=15\mathrm{dB}$.

tests have higher PDs than the 1S-GLRT and 2S-GLRT. Figure 6.7 gives the detection thresholds of the detectors under different σ_n^2 and ε. It can be seen that the detection thresholds remain almost the same as the increase of σ_n^2 and ε.

6.3 ADAPTIVE DETECTORS IN SUBSPACE INTERFERENCE PLUS PARTIALLY HOMOGENEOUS CLUTTER

In this section, we consider the problem of detecting a distributed target in subspace interference plus partially homogeneous clutter. The main difference with Section 6.2 is that the clutter is partially homogeneous. Hence, the detection problem can also be described as it appears in Equation 6.59. However, in a PHE, we have $\boldsymbol{R}_t=\sigma^2\boldsymbol{R}$, where σ^2 denotes the power mismatch between the clutter in test and training data.

6.3.1 Design of the Detector

We propose two detectors according to the 2S-GLRT and 2S-Wald test[2], which contains two steps. The first step is to derive the GLRT or Wald test with $\boldsymbol{R}$ fixed. The other step is to substitute a proper estimator for $\boldsymbol{R}$.

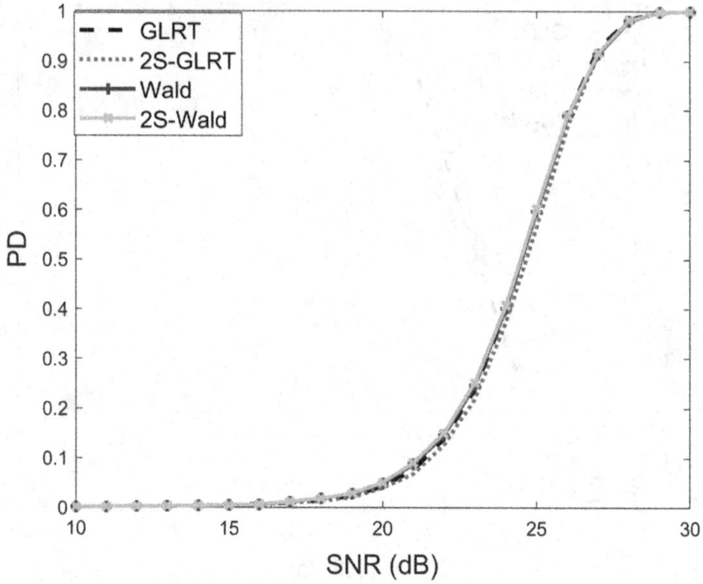

FIGURE 6.5 PDs of the detectors under different SNRs. $p=2$, $q=3$, $K=6$, $L=32$, $\sigma_n^2=1$, $\varepsilon=0.95$, and INR = 15dB.

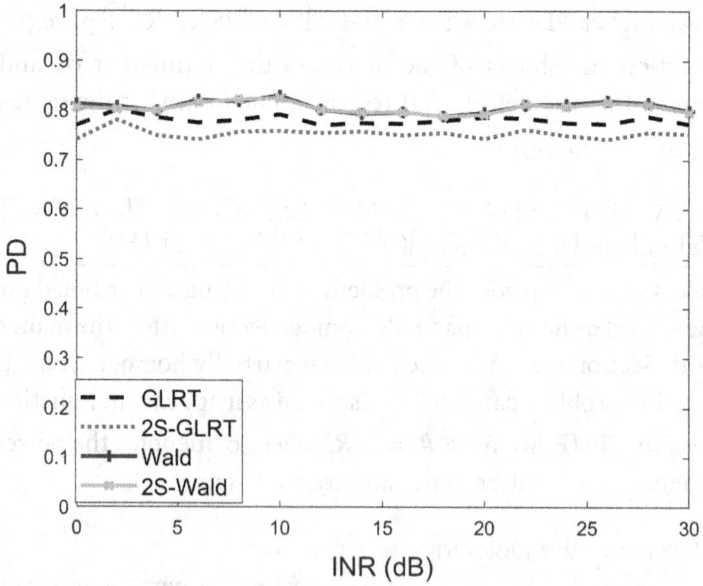

FIGURE 6.6 PDs of the detectors under different INRs. $p=2, q=3, K=6, L=32$, $\sigma_n^2=1$, $\varepsilon=0.95$, and SNR = 26dB

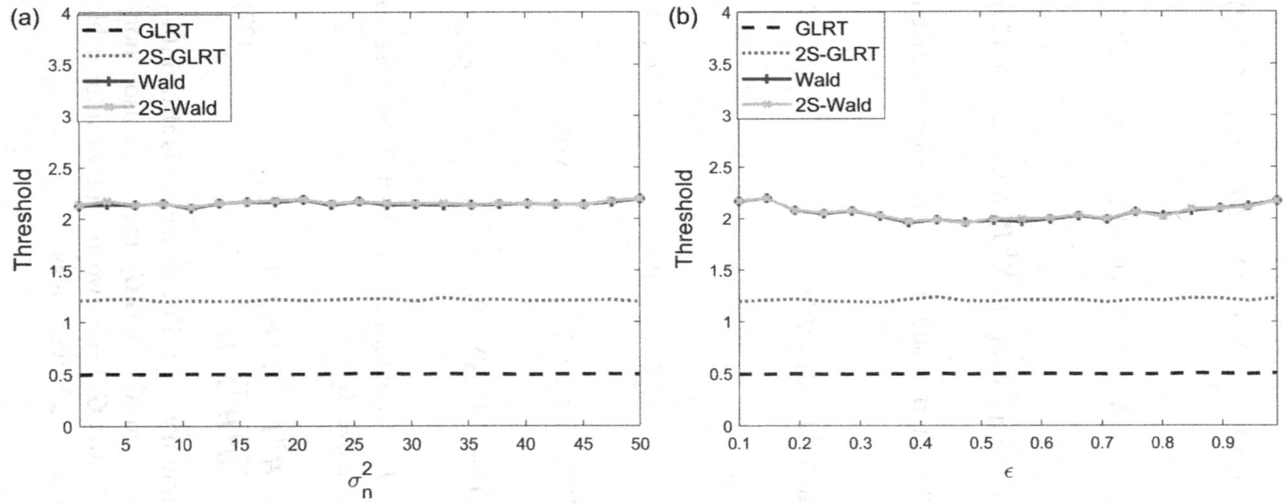

FIGURE 6.7 Detection thresholds of the detectors under different σ_n^2 and ε. $p=2$, $q=3$, $K=6$, and $L=32$. (a) Different ε. $\sigma_n^2=1$. (b) Different σ_n^2. $\varepsilon=0.95$.

6.3.1.1 2S-GLRT

The PDF of $\boldsymbol{X}$ under H_1 is

$$f_1(\boldsymbol{X}) = \sigma^{-2NK} c_{\boldsymbol{R}} \exp\{-\mathrm{tr}[(\boldsymbol{X}-\boldsymbol{DE})^H \boldsymbol{R}^{-1}(\boldsymbol{X}-\boldsymbol{DE})]/\sigma^2\}, \quad (6.122)$$

where $c_{\boldsymbol{R}} = [\pi^N \det(\boldsymbol{R})]^{-K}$.

The MLE of $\boldsymbol{E}$ under H_1 for given $\boldsymbol{R}$ is given in (6.103). Plugging (6.103) into (6.122) leads to

$$f_1(\boldsymbol{X};\boldsymbol{E}_1) = [(\pi\sigma^2)^N \det(\boldsymbol{R})]^{-K} \exp[-\mathrm{tr}(\bar{\boldsymbol{X}}^H \boldsymbol{P}_{\bar{\boldsymbol{D}}}^{\perp} \bar{\boldsymbol{X}})/\sigma^2]. \quad (6.123)$$

Nulling the derivative of (6.123) w.r.t. σ^2 results in the MLE of σ^2 for given $\boldsymbol{R}$ under H_1

$$\hat{\sigma}_1^2 = \mathrm{tr}(\bar{\boldsymbol{X}}^H \boldsymbol{P}_{\bar{\boldsymbol{D}}}^{\perp} \bar{\boldsymbol{X}})/(NK). \quad (6.124)$$

Taking (6.124) into (6.123) yields

$$f_1(\boldsymbol{X};\boldsymbol{E}_1,\sigma_1^2) = c_{\boldsymbol{R}}'[\mathrm{tr}(\bar{\boldsymbol{X}}^H \boldsymbol{P}_{\bar{\boldsymbol{D}}}^{\perp} \bar{\boldsymbol{X}})]^{-NK}, \quad (6.125)$$

where $c_{\boldsymbol{R}}' = [NK/(e\pi)]^{NK} \det(\boldsymbol{R})^{-K}$. Using (6.107), we can rewrite (6.125) as

$$f_1\left(\boldsymbol{X};\hat{\boldsymbol{E}}_1,\hat{\sigma}_1^2\right) = c_{\boldsymbol{R}}'\left[\mathrm{tr}\left(\bar{\boldsymbol{X}}^H \boldsymbol{P}_{\bar{\boldsymbol{J}}}^{\perp} \bar{\boldsymbol{D}}\right) - \bar{\boldsymbol{s}}^H \boldsymbol{P}_{\bar{\boldsymbol{J}}}^{\perp} \bar{\boldsymbol{X}}\bar{\boldsymbol{X}}^H \boldsymbol{P}_{\bar{\boldsymbol{J}}}^{\perp} \bar{\boldsymbol{s}} / \bar{\boldsymbol{s}}^H \boldsymbol{P}_{\bar{\boldsymbol{J}}}^{\perp} \bar{\boldsymbol{s}}\right]^{-NK}, \quad (6.126)$$

where $\bar{\boldsymbol{s}} = \bar{\boldsymbol{H}}\boldsymbol{\theta}$ with $\bar{\boldsymbol{H}} = \boldsymbol{R}^{-1/2}\boldsymbol{H}$. To obtain the MLE of θ, we need to maximize the following function over θ

$$h(\boldsymbol{\theta}) = \frac{\boldsymbol{\theta}^H \bar{\boldsymbol{H}}^H \boldsymbol{P}_{\bar{\boldsymbol{J}}}^{\perp} \bar{\boldsymbol{X}}\bar{\boldsymbol{X}}^H \boldsymbol{P}_{\bar{\boldsymbol{J}}}^{\perp} \bar{\boldsymbol{H}}\boldsymbol{\theta}}{\boldsymbol{\theta}^H \bar{\boldsymbol{H}}^H \boldsymbol{P}_{\bar{\boldsymbol{J}}}^{\perp} \bar{\boldsymbol{H}}\boldsymbol{\theta}}. \quad (6.127)$$

It is easy to show that the maximum value of Equation 6.127 is also (6.109).

Note that if λ_* is a non-zero eigenvalue of $\boldsymbol{G}_1\boldsymbol{G}_2$, then λ_* is also a non-zero eigenvalue of $\boldsymbol{G}_2\boldsymbol{G}_1$, with $\boldsymbol{G}_1$ and $\boldsymbol{G}_2$ being two arbitrary conformable matrices [23]. This fact is mathematically written as

$$\lambda_*(\boldsymbol{G}_1\boldsymbol{G}_2) = \lambda_*(\boldsymbol{G}_2\boldsymbol{G}_1) \quad (6.128)$$

According to (6.128), we have

$$\lambda_{\max}(\boldsymbol{K}_{\boldsymbol{R}}) = \lambda_{\max}(\bar{\boldsymbol{X}}^H \boldsymbol{P}_{\boldsymbol{P}_{\bar{J}}^{\perp}\bar{\boldsymbol{H}}} \bar{\boldsymbol{X}}), \tag{6.129}$$

where $\boldsymbol{K}_{\boldsymbol{R}}$ is given in (6.110) and

$$\boldsymbol{P}_{\boldsymbol{P}_{\bar{J}}^{\perp}\bar{\boldsymbol{H}}} = \boldsymbol{P}_{\bar{J}}^{\perp}\bar{\boldsymbol{H}}(\bar{\boldsymbol{H}}^H \boldsymbol{P}_{\bar{J}}^{\perp}\bar{\boldsymbol{H}})^{-1}\bar{\boldsymbol{H}}^H \boldsymbol{P}_{\bar{J}}^{\perp}. \tag{6.130}$$

Gathering the results above leads to

$$f_1(\boldsymbol{X}; \hat{\boldsymbol{E}}_1, \hat{\sigma}_1^2, \hat{\theta}_1) = c_{\boldsymbol{R}}'[\mathrm{tr}(\bar{\boldsymbol{X}}^H \boldsymbol{P}_{\bar{J}}^{\perp}\bar{\boldsymbol{X}}) - \lambda_{\max}(\bar{\boldsymbol{X}}^H \boldsymbol{P}_{\boldsymbol{P}_{\bar{J}}^{\perp}\bar{\boldsymbol{H}}}\bar{\boldsymbol{X}})]^{-NK}. \tag{6.131}$$

In a similar manner to (6.125), we have

$$f_0(\boldsymbol{X}; \hat{\boldsymbol{Q}}_0, \hat{\sigma}_0^2) = c_{\boldsymbol{R}}'[\mathrm{tr}(\bar{\boldsymbol{X}}^H \boldsymbol{P}_{\bar{J}}^{\perp}\bar{\boldsymbol{X}})]^{-NK}. \tag{6.132}$$

Taking the (NK)th root of the ratio of (6.131) to (6.132) results in the GLRT for given $\boldsymbol{R}$ as

$$t_{GLRT|\boldsymbol{R}} = \frac{\mathrm{tr}(\bar{\boldsymbol{X}}^H \boldsymbol{P}_{\bar{J}}^{\perp}\bar{\boldsymbol{X}})}{\mathrm{tr}(\bar{\boldsymbol{X}}^H \boldsymbol{P}_{\bar{J}}^{\perp}\bar{\boldsymbol{X}}) - \lambda_{\max}(\bar{\boldsymbol{X}}^H \boldsymbol{P}_{\boldsymbol{P}_{\bar{J}}^{\perp}\bar{\boldsymbol{H}}}\bar{\boldsymbol{X}})}. \tag{6.133}$$

The MLE of $\boldsymbol{R}$ based on the training data set is $\hat{\boldsymbol{R}} = \boldsymbol{S}/L$. Substituting $\hat{\boldsymbol{R}}$ into (6.133) yields the final 2S-GLRT

$$t_{NAMDD\text{-}IR} = \frac{\mathrm{tr}(\tilde{\boldsymbol{X}}^H \boldsymbol{P}_{\tilde{J}}^{\perp}\tilde{\boldsymbol{X}})}{\mathrm{tr}(\tilde{\boldsymbol{X}}^H \boldsymbol{P}_{\tilde{J}}^{\perp}\tilde{\boldsymbol{X}}) - \lambda_{\max}(\tilde{\boldsymbol{X}}^H \boldsymbol{P}_{\boldsymbol{P}_{\tilde{J}}^{\perp}\tilde{\boldsymbol{H}}}\tilde{\boldsymbol{X}})}, \tag{6.134}$$

which is statistically equivalent to

$$t_{NAMDD\text{-}IR}' = \frac{\lambda_{\max}(\tilde{\boldsymbol{X}}^H \boldsymbol{P}_{\boldsymbol{P}_{\tilde{J}}^{\perp}\tilde{\boldsymbol{H}}}\tilde{\boldsymbol{X}})}{\mathrm{tr}(\tilde{\boldsymbol{X}}^H \boldsymbol{P}_{\tilde{J}}^{\perp}\tilde{\boldsymbol{X}})}. \tag{6.135}$$

For convenience, the detector in (6.134) or (6.135) is referred to as the normalized adaptive matched direction detector with interference rejection (NAMDD-IR).

It is worth noting that the NAMDD-IR possesses the CFAR property w.r.t. $\boldsymbol{R}$ and σ^2. To prove this fact, it is sufficient to show the statistical properties of the quantities $\tilde{\boldsymbol{X}}^H \boldsymbol{P}_{\boldsymbol{P}_{\tilde{\boldsymbol{J}}}^{\perp}\tilde{\boldsymbol{H}}} \tilde{\boldsymbol{X}}$ and $\tilde{\boldsymbol{X}}^H \boldsymbol{P}_{\tilde{\boldsymbol{J}}}^{\perp} \tilde{\boldsymbol{X}}$ under H_0 are independent of $\boldsymbol{R}$ and σ^2. This can be achieved in a manner similar to that in [24]. For brevity, the detailed derivation is omitted.

6.3.1.2 2S-Wald Test

For the 2S-Wald test in the PHE, we have $\Theta_r = \alpha$ and $\Theta_s = [vec^T(\mathbf{Q}), \sigma^2]^T$. Taking the logarithm of Equation 6.122 and performing the derivative w.r.t. $\boldsymbol{\theta}^*$ and $\boldsymbol{\theta}^T$ yields

$$\frac{\partial \ln f_1(\boldsymbol{X}, \boldsymbol{X}_L)}{\partial \boldsymbol{\theta}^*} = \frac{1}{\sigma^2} \boldsymbol{H}^H \boldsymbol{R}^{-1} \boldsymbol{X}_1 \alpha \tag{6.136}$$

and

$$\frac{\partial \ln f_1(\boldsymbol{X}, \boldsymbol{X}_L)}{\partial \boldsymbol{\theta}^T} = \frac{1}{\sigma^2} \alpha^H \boldsymbol{X}_1^H \boldsymbol{R}^{-1} \boldsymbol{H} \tag{6.137}$$

Then, similar to (6.72), we can obtain

$$\boldsymbol{J}_{\Theta_r,\Theta_r}(\Theta) = \frac{1}{\sigma^2} \alpha^H \alpha \cdot \boldsymbol{H}^H \boldsymbol{R}^{-1} \boldsymbol{H}. \tag{6.138}$$

Substituting (6.136), (6.137), and (6.138) into (6.63) results in

$$t_{Wald|\sigma^2,\boldsymbol{\theta},\alpha,\boldsymbol{R}} = \frac{\alpha^H \alpha}{\sigma^2} \cdot \boldsymbol{\theta}^H \boldsymbol{H}^H \boldsymbol{R}^{-1} \boldsymbol{H} \boldsymbol{\theta}, \tag{6.139}$$

The MLEs of α and $\boldsymbol{\theta}$ given in (6.104) and (6.109), respectively. Substituting (6.104) and (6.124) into (6.139) and dropping the constant results in the Wald test for given $\boldsymbol{\theta}$ and $\boldsymbol{R}$ as

$$t_{Wald|\,\boldsymbol{\theta},\boldsymbol{R}} = \frac{\bar{\boldsymbol{s}}^H \boldsymbol{P}_{\bar{\boldsymbol{J}}}^{\perp} \bar{\boldsymbol{X}} \bar{\boldsymbol{X}}^H \boldsymbol{P}_{\bar{\boldsymbol{J}}}^{\perp} \bar{\boldsymbol{H}}_{,\ K_R} \cdot \bar{\boldsymbol{s}}^H \bar{\boldsymbol{s}}}{(\bar{\boldsymbol{s}}^H \boldsymbol{P}_{\bar{\boldsymbol{J}}}^{\perp} \bar{\boldsymbol{s}})^2 \operatorname{tr}(\bar{\boldsymbol{X}}^H \boldsymbol{P}_{\bar{\boldsymbol{D}}}^{\perp} \bar{\boldsymbol{X}})}. \tag{6.140}$$

Substituting the sample covariance matrix for the implicitly $\bar{\boldsymbol{R}}$ used in (6.140) results in the final 2S-Wald test

$$t_{NOMDD\text{-}IR} = \frac{\operatorname{tr}(\tilde{\boldsymbol{X}}^H \boldsymbol{P}_{\tilde{\boldsymbol{s}}_K|\tilde{\boldsymbol{J}}}^H \boldsymbol{P}_{\tilde{\boldsymbol{s}}_K|\tilde{\boldsymbol{J}}} \tilde{\boldsymbol{X}})}{\operatorname{tr}(\tilde{\boldsymbol{X}}^H \boldsymbol{P}_{\tilde{\boldsymbol{D}}}^{\perp} \tilde{\boldsymbol{X}})}. \tag{6.141}$$

which is referred to as the normalized oblique matched direction detector with interference rejection (NOMDD-IR). In (6.141), $\tilde{s}_K$ is given in Equation 6.115, and $\tilde{\boldsymbol{D}}=[\tilde{\boldsymbol{s}}_K,\tilde{\boldsymbol{J}}]$.

6.3.2 Numerical Examples

In this subsection, we evaluate the detection performance of the proposed detectors by Monte Carlo simulations. The SNR is defined as

$$SNR=\frac{1}{\sigma^2}\alpha^H\alpha\theta^H\boldsymbol{H}^H\boldsymbol{R}^{-1}\boldsymbol{H}\theta\,, \tag{6.142}$$

while the INR is defined as

$$INR=\frac{1}{\sigma^2}\mathrm{tr}(\boldsymbol{Q}^H\boldsymbol{J}^H\boldsymbol{R}^{-1}\boldsymbol{J}\boldsymbol{Q})\,. \tag{6.143}$$

For comparison purposes, we also plot the PDs of the following two detectors, proposed in [4]

$$t_{GLRT}=\frac{(\hat{\sigma}_0^2)^{NK/(K+L)}\det(\boldsymbol{I}_K+\tilde{\boldsymbol{X}}^H\boldsymbol{P}_{\tilde{J}}^{\perp}\tilde{\boldsymbol{X}}/\hat{\sigma}_0^2)}{(\hat{\sigma}_1^2)^{NK/(K+L)}\det(\boldsymbol{I}_K+\tilde{\boldsymbol{X}}^H\boldsymbol{P}_{\tilde{B}}^{\perp}\tilde{\boldsymbol{X}}/\hat{\sigma}_1^2)}, \tag{6.144}$$

$$t_{2S\text{-}GLRT}=\frac{\mathrm{tr}(\tilde{\boldsymbol{X}}^H\boldsymbol{P}_{\tilde{J}}^{\perp}\tilde{\boldsymbol{X}})}{\mathrm{tr}(\tilde{\boldsymbol{X}}^H\boldsymbol{P}_{\tilde{B}}^{\perp}\tilde{\boldsymbol{X}})}\,. \tag{6.145}$$

The two detectors can work for the detection problem in Equation 6.59, and they have the CFAR property. In Equation 6.144, $\tilde{\boldsymbol{B}}=[\tilde{\boldsymbol{H}},\tilde{\boldsymbol{J}}]$, $\hat{\sigma}_i^2$, $i=0,1$, is the sole solution to the equation

$$\frac{NK}{K+L}-\sum_{k=1}^{t}\frac{\lambda_{k,i}}{\lambda_{k,i}+x}=0, \tag{6.146}$$

where x denotes the unknown, t is the minimum value of N and K, $\lambda_{k,0}$ is the kth non-zero eigenvalue of $\tilde{\boldsymbol{X}}^H\boldsymbol{P}_{\tilde{J}}^{\perp}\tilde{\boldsymbol{X}}$, and $\lambda_{k,1}$ is the kth non-zero eigenvalue of $\tilde{\boldsymbol{X}}^H\boldsymbol{P}_{\tilde{B}}^{\perp}\tilde{\boldsymbol{X}}$.

We choose $\sigma^2=1$, and other parameters are the same as those in Section 6.2.

Figures 6.8 and 6.9 show the PDs of the detectors versus SNR with different fixed values of q. Comparing the results reveals that when p and q

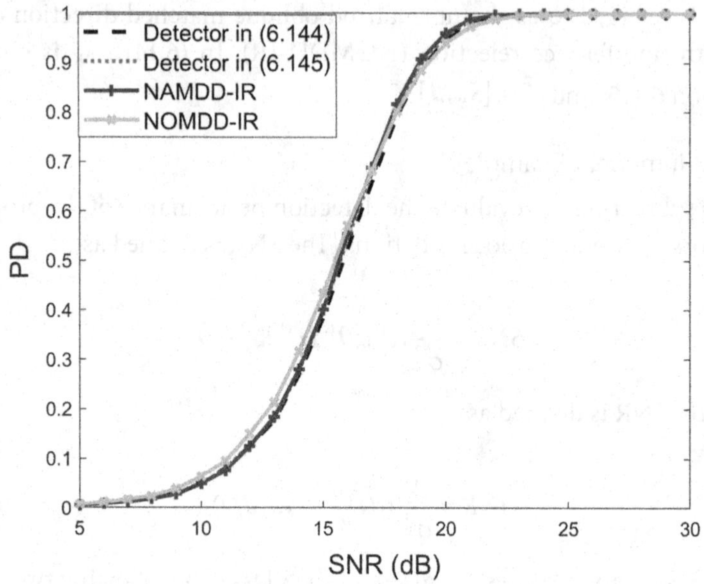

FIGURE 6.8 PD versus SNR. $L = 2N$, $p = 2$, $q = 1$, and $K = 2$.

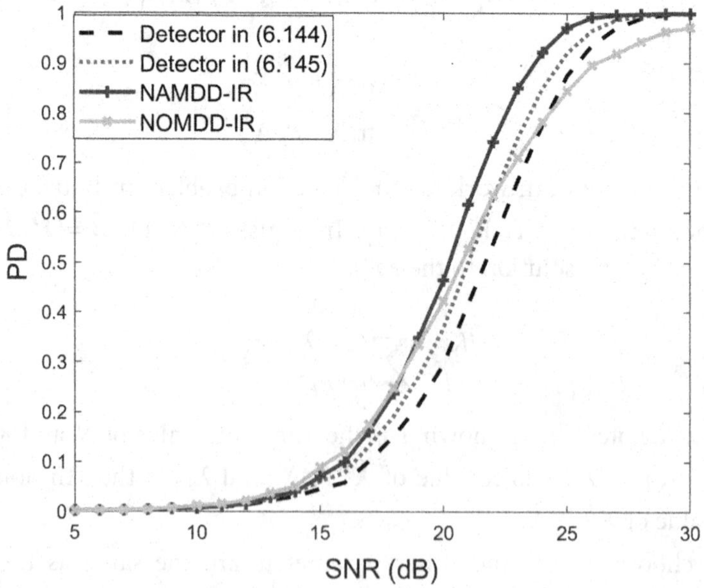

FIGURE 6.9 PD versus SNR. $L = 2N$, $p = 2$, $q = 4$, and $K = 2$.

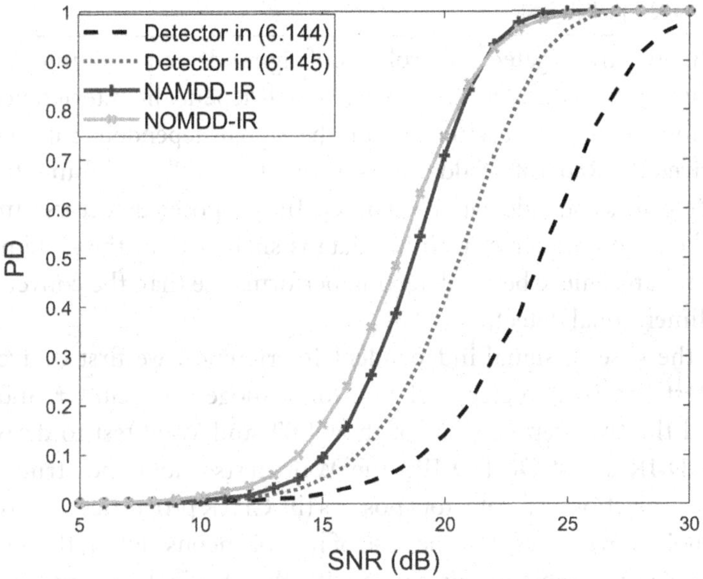

FIGURE 6.10 PD versus SNR. $L=2N$, $p=4$, $q=2$, and $K=4$.

are both small, the NAMDD-IR, NOMDD-IR, and detectors in (6.144) and (6.145) nearly have the same detection performance. In contrast, when q is large the NAMDD-IR achieves the best detection performance. Moreover, the PDs of all the detectors decrease as $q$increases. This is owing to the increase in the loss of the signal energy projected in the interference subspace.

It is shown in Figure 6.10 that when p is large, both the NAMDD-IR and NOMDD-IR provide significantly better detection performance than the detectors in Equations 6.144 and 6.145. This is because with the increase of p, the uncertainty in the signal steering vector $\boldsymbol{s}$ becomes serious, and the NAMDD-IR and NOMDD-IR take into account this uncertainty, whereas the detectors in Equations 6.144 and 6.145 do not. Particularly, the performance improvement of the NAMDD-IR, in terms of SNR, w.r.t. the detector in Equation 6.144 at PD=0.9 is more than 2 dB.

It is worth pointing out that the PDs of some detectors increase with the same SNR in Figure 6.10, compared with the results in Figure 6.9. One possible reason is that the angle between the whitened signal and the whitened interference subspace is smaller in Figure 6.9 than that in Figure 6.10. This angle plays a key role in controlling the detection performance of the detectors [25].

6.4 CONCLUSION

This chapter investigated the problem of signal detection in interference. Two cases were considered, namely, signal-dependent interference and signal-independent interference. For the signal-dependent interference, we derived the Rao test, Wald test, and two-step GLRT, and found that the three detectors coincide with each other. The proposed detectors can work well when the number of training data is smaller than the system DOF and they can achieve better detection performance than the conventional multidimensional detectors.

For the case of signal-independent interference, we first derived the Wald test and its two-step variation for homogeneous clutter, and then we used the two-step versions of the GLRT and Wald test to derive the NAMDD-IR and NOMDD-IR. The Wald tests adopt the structure of oblique projection. All detectors possess the CFAR properties with respect to the noise covariance. For the case of homogeneous clutter, the 1S-Wald test has a comparable performance with the 2S-Wald test, and in some parameter setting the 1S-Wald test and 2S-Wald test can provide slightly better performance than the 1S-GLRT and 2S-GLRT. For the case of partially homogeneous clutter, the NAMDD-IR and NOMDD-IR can provide better detection performance than existing GLRT and 2S-GLRT detectors and the NAMDD-IR has the best detection performance.

APPENDIX 6.A THE DERIVATION OF EQUATIONS 6.147 AND 6.148

In this appendix, we show Equations 6.117 and 6.118 can be recast as

$$t_{GLRT} = \lambda\left[\tilde{\boldsymbol{X}}^H \boldsymbol{P}_{\boldsymbol{P}_{\tilde{J}}^{\perp}\tilde{\boldsymbol{H}}} \tilde{\boldsymbol{X}}(\boldsymbol{I}_K + \tilde{\boldsymbol{X}}^H \boldsymbol{P}_{\tilde{J}}^{\perp} \tilde{\boldsymbol{X}})^{-1}\right] \tag{6.147}$$

and

$$t_{2S\text{-}GLRT} = \lambda\left(\tilde{\boldsymbol{X}}^H \boldsymbol{P}_{\boldsymbol{P}_{\tilde{J}}^{\perp}\tilde{\boldsymbol{H}}} \tilde{\boldsymbol{X}}\right) \tag{6.148}$$

respectively.

Using (6.128), we can rewrite (6.117) as

$$t_{GLRT} = \lambda\left[\tilde{\boldsymbol{X}}(\boldsymbol{I}_K + \tilde{\boldsymbol{X}}^H \boldsymbol{P}_{\tilde{J}}^{\perp} \tilde{\boldsymbol{X}})^{-1} \tilde{\boldsymbol{x}}^H \boldsymbol{Z}\boldsymbol{P}_{\boldsymbol{Z}^H\tilde{\boldsymbol{H}}} \boldsymbol{Z}^H\right] \tag{6.149}$$

Substituting $\boldsymbol{P}_{\boldsymbol{Z}^H\tilde{\boldsymbol{H}}} = \boldsymbol{Z}^H\tilde{\boldsymbol{H}}(\tilde{\boldsymbol{H}}^H\boldsymbol{P}_{\tilde{\boldsymbol{J}}}^{\perp}\tilde{\boldsymbol{H}})^{-1}\tilde{\boldsymbol{H}}^H\boldsymbol{Z}$ into $\boldsymbol{Z}\boldsymbol{P}_{\boldsymbol{Z}^H\tilde{\boldsymbol{H}}}\boldsymbol{Z}^H$ and using the equality $\boldsymbol{P}_{\tilde{\boldsymbol{J}}}^{\perp} = \boldsymbol{Z}\boldsymbol{Z}^H$, we arrive at

$$\boldsymbol{Z}\boldsymbol{P}_{\boldsymbol{Z}^H\tilde{\boldsymbol{H}}}\boldsymbol{Z}^H = \boldsymbol{P}_{\boldsymbol{P}_{\tilde{\boldsymbol{J}}}^{\perp}\tilde{\boldsymbol{H}}} \tag{6.150}$$

Substituting (6.150) into (6.149) leads to

$$t_{GLRT} = \lambda\left[\tilde{\boldsymbol{X}}(\boldsymbol{I}_K + \tilde{\boldsymbol{X}}^H\boldsymbol{P}_{\tilde{\boldsymbol{J}}}^{\perp}\tilde{\boldsymbol{X}})^{-1}\tilde{\boldsymbol{X}}^H\boldsymbol{P}_{\boldsymbol{P}_{\tilde{\boldsymbol{J}}}^{\perp}\tilde{\boldsymbol{H}}}\right] \tag{6.151}$$

which can be recast as (6.147), according to (6.128).

In a similar manner, we can represent (6.118) as (6.148).

NOTES

1. The GLRT and 2S-GLRT can be written as the forms directly with the original data, which are given in the Appendix 6.A.
2. We do not consider one-step GLRT criterion due to the following two facts. First and foremost, the one-step criterion is not guaranteed to have superior detection performance to the two-step one [22]. The other is that the derivation according to the one-step criterion is rather complicated for the problem at hand, and it does not admit a closed-form solution, except the degenerate case of $K = 1$.

REFERENCES

1. L. Yan, P. Addabbo, C. Hao, D. Orlando, and A. Farina, "New ECCM Techniques Against Noiselike and/or Coherent Interferers," *IEEE Transactions on Aerospace and Electronic Systems*, vol. 56, no. 2, pp. 1172–1188, 2020.
2. D. Orlando, "A Novel Noise Jamming Detection Algorithm for Radar Applications," *IEEE Signal Processing Letters*, vol. 24, no. 2, pp. 206–210, 2017.
3. M. Hurtado and A. Nehorai, "Polarimetric Detection of Targets in Heavy Inhomogeneous Clutter," *IEEE Transactions on Signal Processing*, vol. 56, no. 4, pp. 1349–1361, 2008.
4. F. Bandiera, A. De Maio, A. S. Greco, and G. Ricci, "Adaptive Radar Detection of Distributed Targets in Homogeneous and Partially Homogeneous Noise Plus Subspace Interference," *IEEE Transactions on Signal Processing*, vol. 55, no. 4, pp. 1223–1237, 2007.
5. M. Sun, W. Liu, J. Liu, and C. Hao, "Multichannel Adaptive Detection Based on Gradient Test and Durbin Test in Deterministic Interference and Structure Nonhomogeneity," *IEEE Signal Processing Letters*, vol. 29, pp. 592–596, 2022.

6. O. Bialer, D. Raphaeli, and A. J. Weiss, "Maximum-Likelihood Direct Position Estimation in Dense Multipath," *IEEE Transactions on Vehicular Technology*, vol. 62, no. 5, pp. 2069–2079, 2013.
7. E. J. Kelly and K. M. Forsythe, *Adaptive Detection and Parameter Estimation for Multidimensional Signal Models*, Lexington: Lincoln Laboratory, Tech. Rep. 848, 1989.
8. W. Liu, Y. Wang, and W. Xie, "Fisher Information Matrix, Rao Test, and Wald Test for Complex-Valued Signals and Their Applications," *Signal Processing*, vol. 94, pp. 1–5, 2014.
9. S. M. Kay, *Fundamentals of Statistical Signal Processing, Volume I: Estimation Theory*. Prentice-Hall, Englewood Cliffs, NJ, 1998.
10. A. Hjørungnes, *Complex-Valued Matrix Derivatives: with Applications in Signal Processing and Communications*. Cambridge University Press, Cambridge, 2011.
11. J. Sherman and W. J. Morrison, "Adjustment of an Inverse Matrix Corresponding to a Change in One Element of a Given Matrix," *The Annals of Mathematical Statistics*, vol. 21, no. 1, pp. 124–127, 1950.
12. T. W. Anderson, *An Introduction to Multivariate Statistical Analysis*. Wiley, Hoboken, NJ, 2003.
13. W. Liu, W. Xie, J. Liu, and Y. Wang, "Adaptive Double Subspace Signal Detection in Gaussian Background-Part I: Homogeneous Environments," *IEEE Transactions on Signal Processing*, vol. 62, no. 9, pp. 2345–2357, 2014.
14. H.-R. Park, J. Li, and H. Wang, "Polarization-Space-Time Domain Generalized Likelihood Ratio Detection of Radar Targets," *Signal Processing*, vol. 41, no. 2, pp. 153–164, 1995.
15. J. Liu, W. Liu, B. Chen, H. Liu, H. Li, and C. Hao, "Modified Rao Test for Multichannel Adaptive Signal Detection," *IEEE Transactions on Signal Processing*, vol. 64, no. 3, pp. 714–725, 2016.
16. M. Hurtado, J. J. Xiao, and A. Nehorai, "Target Estimation, Detection, and Tracking a Look at Adaptive Polarimetric Design," *IEEE Signal Processing Magazine*, vol. 26, no. 1, pp. 42–52, 2009.
17. W. Liu, J. Li, P. Wang, X. Zha, and Y.-L. Wang, "Wald Tests for Signal Detection When Uncertainty Exists in A Target's Spatial-Temporal Steering Vector," *Science China Information Sciences*, vol. 63, no. 8, pp. 1–3, 2020.
18. N. Li, G. Cui, H. Yang, L. Kong, Q. H. Liu, and S. Iommelli, "Adaptive Detection of Moving Target with MIMO Radar in Heterogeneous Environments Based on Rao and Wald Tests," *Signal Processing*, vol. 114, pp. 198–208, 2015.
19. H. Yanai, K. Takeuchi, and Y. Takane, *Projection Matrices, Generalized Inverse Matrices, And Singular Value Decomposition (Statistics for Social and Behavioral Sciences)*, Springer, New York, 2011.
20. H. L. V. Trees, *Detection, Estimation, And Modulation Theory, Part IV: Optimum Array Processing*, Wiley, New York, 2002.
21. F. Bandiera, O. Besson, D. Orlando, G. Ricci, and L. L. Scharf, "GLRT-Based Direction Detectors in Homogeneous Noise and Subspace Interference," *IEEE Transactions on Signal Processing*, vol. 55, no. 6, pp. 2386–2394, 2007.

22. O. Besson, L. L. Scharf, and S. Kraut, "Adaptive Detection of a Signal Known Only to Lie on A Line in a Known Subspace, When Primary and Secondary data Are Partially Homogeneous," *IEEE Transactions on Signal Processing*, vol. 54, no. 12, pp. 4698–4705, 2006.
23. W. Liu, W. Xie, J. Liu, D. Zou, H. Wang, and Y.-L. Wang, "Detection of A Distributed Target with Direction Uncertainty," *IET Radar, Sonar and Navigation*, vol. 8, no. 9, pp. 1177–1183, 2014.
24. F. Bandiera, D. Orlando, and G. Ricci, "On the CFAR Property of GLRT-based Direction Detectors," *IEEE Transactions on Signal Processing*, vol. 55, no. 8, pp. 4312–4315, 2007.
25. W. Liu, J. Liu, L. Huang, C. Hao, and Y.-L. Wang, "Performance Analysis of Adaptive Detectors for Point Targets in Subspace Interference and Gaussian Noise," *IEEE Transactions on Aerospace and Electronic Systems*, vol. 54, no. 1, pp. 429–441, 2018.

CHAPTER 7

Future Trends

IN THE PREVIOUS CHAPTERS, we introduced the framework and design criteria of adaptive detection methods and discussed adaptive detection algorithms in three types of non-ideal scenarios: adaptive detection algorithms in a sample-starved environment, adaptive detection algorithms in signal mismatch scenarios, and adaptive detection algorithms in interference. However, there are still many problems in adaptive detection that need to be further studied.

In Chapter 2, the problem of detecting targets in a sample-starved environment is discussed. The prior information-based methods and the dimension reduction method are designed to reduce the requirement of sufficient IID training data. For prior information-based methods, the persymmetric structure of the noise covariance matrix and the AR model are exploited. For the dimension reduction method, a novel detection scheme is proposed by projecting the data into several orthogonal subspaces. Numerical results show that the persymmetric detectors and the reduced-dimension detector achieve good detection performance when the number of training data is smaller than the system freedom. Meanwhile, the adaptive detectors in AR noise can work well when the number of the training data is only one. However, when the environment is extremely heterogeneous, no IID training data will be available. Thus, further investigation of the adaptive detectors without training data is necessary.

From Chapter 3 to Chapter 5, the problem of detecting targets in the presence of signal mismatch is dealt with. Adaptive mismatch selective detectors, adaptive mismatch robust detectors, and tunable adaptive

DOI: 10.1201/9781003477907-7

detectors are designed according to the requirements of different scenarios. The mismatch robust detector in Chapter 4 is designed by introducing a random perturbation whose covariance matrix is proportional to the noise covariance matrix. The investigation of further possible structures for the covariance matrix of the random perturbation and the derivation of adaptive detectors for the general case are necessary. Moreover, the adaptive detectors are mainly designed in Gaussian noise and CG noise when the signal mismatch occurs. Although the Gaussian model is the most widely used noise model, the practical noise may deviate from the presumed model due to uncertainties of the environment. Therefore, the research of other noise models and the corresponding extension of adaptive mismatch selective detectors, adaptive mismatch robust detectors, and tunable adaptive detectors are needed.

In Chapter 6, the problem of adaptive detection of a subspace signal in interference is considered. Adaptive detectors are designed in the interference wherein the subspace is assumed completely known. However, the prior information on the interference is usually unknown in a practical environment. Therefore, the investigation of the detection performance when the interference mismatch occurs and the investigation of how to construct an accurate interference model are essential. Meanwhile, adaptive detectors are designed under the assumption that only one interference exists in Chapter 6. The extension of the designed detectors to the case wherein more than two types of interferences are present is necessary.

Apart from the further research of adaptive detection algorithms in the three types of non-ideal scenarios, the adaptive detection algorithms based on the information geometry may be also a point for further research. With the increasing complexity of the problems to be dealt with, nonlinear problems have gradually become the difficulties and challenges faced by signal processing technology. Traditional linear methods have been unable to effectively solve nonlinear problems. Information geometry is more suitable to solve complex nonlinear problems since it extends traditional Euclidean spaces to Riemannian manifolds and extends Euclidean distance metrics to Riemannian metrics.

To improve the detection performance of the pulse Doppler radar, Barbaresco et al. [1] propose a CFAR algorithm by exploiting the information geometry theory. Specifically, Toeplitz positive definite matrices are constructed according to the echoes of each range cell first. Then, the Riemannian distance between the Toeplitz positive definite matrix of the data under test and the Riemannian mean of the Toeplitz positive definite matrices of the training data is used to determine whether the target is present or not. In [2], the power spectral density matrices of the received signals are constructed and two Riemannian distances between the power spectral density matrices are used to detect narrow-band sonar signals in noise. In [3], Hermitian positive definite matrices of each range cell are constructed first. Then, the matrix spectral norm is used to design mean matrix estimators. Finally, the spectral norm measure-based matrix CFAR detector is designed to detect targets using pulse Doppler radar with a small bunch of pulses.

More recently, the combination of the adaptive detection algorithm and the information geometry theory has attracted close attention. Taking into consideration that the training data may be contaminated by outliers, Aubry et al. [4] have designed an adaptive two-stage detector. In the first stage, a covariance matrix estimator, which is defined as the geometric barycenter of a set of covariance matrix estimates obtained from the training data, is designed and the generalized inner product based on the covariance matrix estimator is used to select the most homogeneous training data. In the second stage, an AMF, which is derived according to the data under test and the selected training data, is used to detect targets. In a similar way, two-step detectors are designed in PHE in [5]. The covariance matrix estimators based on geometric barycenters, which do not rely on the prior information about the probability distribution of the received data, are used to eliminate outliers, and the adaptive coherence estimate based on the geometric barycenters is designed. The work in [4] has also been extended to the CG noise and an adaptive normalized matched filter that replaces the sample covariance matrix with the geometric barycenters is designed in [6]. When the numberof training data is limited, the knowledge-aided covariance matrix estimators that combine the prior covariance matrix with geometric mean estimators are proposed and an adaptive NMF based on the knowledge-aided covariance matrix estimators is designed in non-Gaussian noise [7].

The research of approaches that combine machine learning algorithms and adaptive detection algorithms may be another possible future research consideration. In adaptive detection algorithms, both the target signal model and the noise model should be constructed before the detectors are designed. Although the adaptive detection algorithms have clear physical interpretations, the construction of the models relies heavily on human experience and prior knowledge. In a complex and changeable environment, the precise models are difficult to obtain. Without precise models, the detection performance of the adaptive detectors degenerates. Machine learning algorithms can automatically learn knowledge and information from task-related data sets and continuously optimize and improve algorithms. When the problem to be solved is too complex for existing mathematical theories to provide an accurate model or even if the model is accurate, the optimization of the problem is highly complex and difficult to achieve, machine learning algorithms can be used. Machine learning technology, which can extract the law of data from existing data and apply the law to unknown data, has been applied in synthetic aperture radar (SAR), radar imaging, signal recognition such as radar emitter recognition, and human behaviour recognition based on millimetre wave radar.

In recent years, machine learning technology is beginning to be applied to radar detection. One of the most important scenarios machine learning technology has been applied to is the detection of floating small targets. In [8], to improve the PD, a modified support vector machine is proposed according to the three discriminative features extracted from the received signals in the time and frequency domains. In [9], six features extracted from radar returns in the time, frequency, and time-frequency domains are combined into a feature vector, and a principal component analysis (PCA)-based anomaly detector is proposed to distinguish the target cells from the clutter-only cells. In [10], eight features are exacted and the K-nearest neighbour (KNN) algorithm and anomaly detection are cooperated to improve detection performance. The deep learning algorithms, which can exact the features automatically, have also been used in the problem of detecting floating small targets. In [11], the time-frequency images are used as the inputs and an enhanced convolutional neural network (CNN) is proposed to detect sea-surface small targets.

More recently, the hybrid approaches that combine adaptive detection algorithms with machine learning algorithms have gradually been gaining attention. The hybrid approaches can obtain advantages from both adaptive detection algorithms and data-driven algorithms. In [12], the raw data which are simulated from a chosen model or the detection statistics of the adaptive detectors like Kelly's GLRT, the AMF, and the ACE are used to construct feature vectors, and the KNN approach is exploited to detect radar signals in the Gaussian background. In [13], the primary data and the training data are mapped into a feature vector and a KNN-based radar detector is proposed for coherent targets in non-Gaussian noise. The simulation results demonstrate the superiority of the detector with respect to natural competitors. Therefore, the investigation of how to introduce machine learning to adaptive detection algorithms and combine the advantages of machine learning and adaptive detection to improve detection performance is a promising future trend.

REFERENCES

1. F. Barbaresco, Innovative Tools for Radar Signal Processing Based on Cartan's Geometry of SPD Matrices and Information Geometry, *IEEE Radar Conference*, pp. 1–6, 2008.
2. K. M. Wong, J. K. Zhang, and H. Jiang, Multi-Sensor Signal Processing on a PSD Matrix Manifold, *Sensor Array and Multichannel Signal Processing Workshop*, 2016.
3. W. Zhao, W. Liu, and M. Jin, "Spectral Norm Based Mean Matrix Estimation and Its Application to Radar Target CFAR Detection," *IEEE Transactions on Signal Processing*, vol. 67, no. 22, pp. 5746–5760, 2019.
4. A. Aubry, A. De Maio, L. Pallotta, and A. Farina, "Covariance Matrix Estimation via Geometric Barycenters and Its Application to Radar Training Data Selection," *IET Radar Sonar Navigation*, vol. 7, no. 6, pp. 600–614, 2013.
5. H. Ye, Y.-L. Wang, W. Liu, J. Liu, and H. Chen, "Adaptive Detection in Partially Homogeneous Environment with Limited Samples Based on Geometric Barycenters," *IEEE Signal Processing Letters*, vol. 29, pp. 2083–2087, 2022.
6. G. Cui, N. Li, L. Pallotta, G. Foglia, and L. Kong, "Geometric Barycenters for Covariance Estimation in Compound-Gaussian Clutter," *IET Radar Sonar Navigation*, vol. 11, no. 3, pp. 404–409, 2017.
7. Z. Shang, K. Huo, W. Liu, Y. Sun, and Y.-L. Wang, "Knowledge-Aided Covariance Estimate via Geometric Mean for Adaptive Detection," *Digital Signal Processing*, vol. 97, pp. 1–13, 2020.
8. Y. Li, P. Xie, Z. Tang, T. Jiang, and P. Qi, "SVM-Based Sea-Surface Small Target Detection: A False-Alarm-Rate-Controllable Approach," *IEEE Geoscience and Remote Sensing Letters*, vol. 16, no. 8, pp. 1225–1229, 2019.
9. T. Gu, "Detection of Small Floating Targets on the Sea Surface Based on Multi-Features and Principal Component Analysis," *IEEE Geoscience and Remote Sensing Letters*, vol. 17, no. 5, pp. 809–813, 2020.
10. Z.-X. Guo and P.-L. Shui, "Anomaly Based Sea-Surface Small Target Detection Using K-Nearest Neighbor Classification," *IEEE Transactions on Aerospace and Electronic Systems*, vol. 56, no. 6, pp. 4947–4964, 2020.
11. X. Chen, N. Su, Y. Huang, and J. Guan, "False-Alarm-Controllable Radar Detection for Marine Target Based on Multi Features Fusion via CNNs," *IEEE Sensors Journal*, vol. 21, no. 7, pp. 9099–9111, 2021.
12. A. Coluccia, A. Fascista, and G. Ricci, "A K-Nearest Neighbors Approach to the Design of Radar Detectors," *Signal Processing*, vol. 174, pp. 1–10, 2020.
13. A. Coluccia, A. Fascista, and G. Ricci, "A KNN-Based Radar Detector for Coherent Targets in Non-Gaussian Noise," *IEEE Signal Processing Letters*, vol. 28, pp. 778–782, 2021.